International Conference on Materials in Civil Engineering (MICE 2020)

AA007351

Held online due to COVID-19

IOP Conference Series: Materials Science and Engineering
Volume 1028

Sanya, China
23 - 25 October 2020

Editor:

Weimin Huang

ISBN: 978-1-7138-3328-4
ISSN: 1757-8981

Printed from e-media with permission by:

Curran Associates, Inc.
57 Morehouse Lane
Red Hook, NY 12571

Some format issues inherent in the e-media version may also appear in this print version.

Institute of Physics
Dirac House, Temple Back
Bristol BS1 6BE UK

Phone: 44 1 17 929 7481
Fax: 44 1 17 920 0979

techtracking@iop.org

Additional copies of this publication are available from:

Curran Associates, Inc.
57 Morehouse Lane
Red Hook, NY 12571 USA
Phone: 845-758-0400
Fax: 845-758-2633
Email: curran@proceedings.com
Web: www.proceedings.com

TABLE OF CONTENTS

IOP Conf. Series: Materials Science and Engineering 1028 (2021) 011001 doi:10.1088/1757-899X/1028/1/011001

Preface

Weimin Huang

Editor, Proceedings of the International Conference on Materials in Civil Engineering (MICE 2020)

Associate Professor, School of Mechanical and Aerospace Engineering, Nanyang Technological University, Singapore

Email: MWMHuang@ntu.edu.sg

The International Conference on Materials in Civil Engineering (MICE 2020) was planned to be held on October 23-25, 2020 in Sanya, China. Due to the COVID-19 epidemic and to avoid the risk of crowd gathering, the organizing committee decided to hold MICE 2020 virtually/online on October 23-25, 2020. Although held online, this year's conference also attracted many scholars and experts in the field of Materials and Civil Engineering to share new research results, novel policy developments, and practical implementation experiences regarding. MICE2020 was a successful, fruitful online academic event. Nearly 40 participants from 4 different countries and regions attended the online meeting.

The conference consists of three parts: keynote presentations, oral presentations and poster presentations and discussions covering a wide range of subjects from materials for structural damage monitoring, potential applications of SMAs in buildings and bridges, recycled waste materials in reinforced concrete construction, study on the properties of concretes at the nanoscale. We are very grateful that Professor Qing Wang from Shandong University of Science and Technology presided over the morning conference, and Professor Mohamed Abdel Kader El Gelany Ismail, from Miami College of Henan University, took charge of the afternoon conference. Thanks for all participants for attending and sharing!

It is our great honor to invite 4 experts to give the outstanding and excellent keynote speeches: Prof. Qing Wang, from Shandong University of Science and Technology, China; Prof. Shahria Alam, from The University of British Columbia (Okanagan Campus), Canada; Prof. Faiz Uddin Ahmed Shaikh, from Curtin University, Australia and Prof. Mohamed Abdel Kader El Gelany Ismail, from Miami College of Henan University.

After keynote presentations, there were 8 orals presentations and 7 poster presentations. Oral speakers and poster presenters shared their latest research results with the audience, the audience actively and enthusiastically participated in by asking questions and voting.

This conference proceeding included 15 accepted articles selected from 41 submissions.

We would like to express our gratitude to the reviewers of these manuscripts, who provided their time and expertise to review papers and facilitate the smooth running of MICE. We are extremely grateful for organizers, technical program committee and editors and extend our most sincere thanks to all the authors for their excellent contribution and work and participants for their active attending. Our sincere thanks also go to the IOP Publishing editors and managers for their helpful cooperation during the preparation of the proceeding.

On behalf of the Organizing Committee of MICE 2020

IOP Conf. Series: Materials Science and Engineering 1028 (2021) 011001 doi:10.1088/1757-899X/1028/1/011001

Technical Program Committee

Conference Chair
Wei Min Huang, Nanyang Technological University, Singapore

Conference Co-chair
Shuhua Liu, Wuhan University, China
Qing Wang, Shandong University of Science and Technology, China

TPC Chair
Kong Fah TEE, University of Greenwich, United Kingdom

TPC member
Shahria Alam, the University of British Columbia (Okanagan Campus), Canada
Wahid Ferdous, University of Southern Queensland, Toowoomba, Australia
Thomas F. George, University of Missouri–St. Louis, USA
Meijie Zhang, Wuhan University of Science and Technology, China
Jamal Khatib, University of Wolverhampton, Faculty of Science and Engineering, United Kingdom
Ali Kaveh, Iran University of Science and Technology, Iran
Abhijit Mukherjee, Curtin University, Australia
Trilok N Singh, Indian Institute of Technology Bombay, India
Lan Wang, Inner Mongolia University of Science & Technology, China
Marek Skakuj, University of Wuppertal, Germany
Yanhuai Ding, Xiangtan University, China
Manuel F. M. Costa, University of Minho, Portugal
Hongxia Qiao, Lanzhou University of Technology, China
Resat Oyguc, Istanbul Technical University, Turkey
Shuqing Hao, China University of Mining and Technology, China
Maochieh Chi, Wufeng University, China
Krzysztof Schabowicz, University of Technology, Poland
Khaled Habib, EBR Center KISR, Kuwait
Chao Chen, Central South University, China
Woubishet Taffese, Aalto University, Finland
Amar Khennane, the University of New South Wales, Sydney
Pornkasem Jongpradist, King Mongkut's University of Technology Thonburi, Thailand
Paulo Cachim, Civil Engineering Department of University of Aveiro, Portugal
José Barroso Aguiar, Department of Civil Engineering of University of Minho, Portugal
Fazia Fouchal, University of Limoges, France
Subhasis Roy, University of Calcutta, Canada
Wangyu Ma, West Texas Paving, Inc, USA
Prasanna Egodawatta, Queensland University of Technology,BRISBANE, Australia
Manoj Khandelwal, Federation University Australia, Australia
Luís Filipe Almeida Bernardo, University of Beira Interior, Portugal
Jui-Sheng Rayson Chou, National Taiwan University of Science and Technology, China
Osman Adiguzel, Firat University, Turkey

IOP Conf. Series: Materials Science and Engineering 1028 (2021) 011001 doi:10.1088/1757-899X/1028/1/011001

Conference Overview

2020 International Conference on Materials in Civil Engineering was held on October 23-25, 2020 via online platform due to the pandemic of COVID-19. It aims to provide an international platform for scholars, practicing engineers, designers, researchers and other practitioners who are interested in all aspects relating to properties and applications of civil engineering materials.

Below is the overview of MICE2020 online conference.

There were nearly 40 experts and scholars from 4 countries and regions attending the MICE2020 online conference. The conference consisted of three parts: keynote presentations, oral presentations and poster presentations. Professor Qing Wang, from Shandong University of Science and Technology, China, presided over the morning conference. Professor Mohamed Abdel Kader El Gelany Ismail, from Miami College of Henan University, took charge of the afternoon conference.

We invited 4 keynote speakers to deliver keynote speeches during the conference.

Figure 1. Keynote Presentation from Qing Wang.

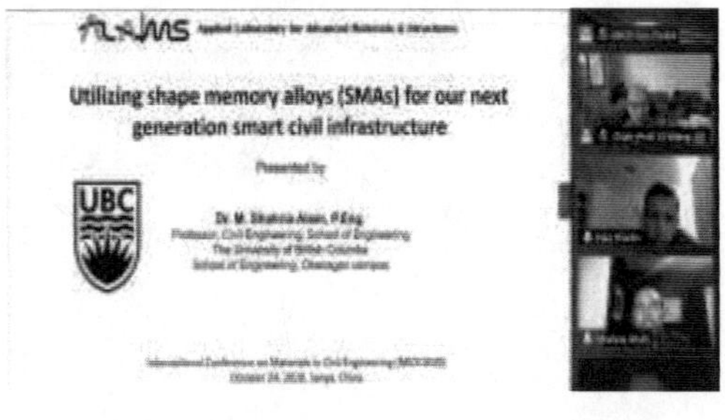

Figure 2. Keynote Presentation from Shahria Alam.

Prof. Qing Wang, the first keynote speaker, representing Shandong University of Science and Technology, China, held a speech on" Study on Flexible Intelligent Sensing Materials for Structural Damage Monitoring"(**Figure 1**).

The second keynote speaker, Prof. Shahria Alam, from The University of British Columbia (Okanagan

Campus), Canada, delivered a speech on" Utilizing shape memory alloys (SMAs) for our next generation smart civil infrastructure"(**Figure 2**).

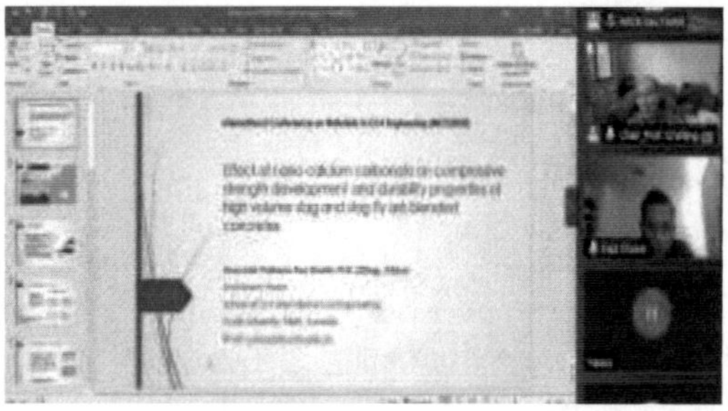

Figure 3. Keynote presentation from Prof. Faiz Uddin Ahmed Shaikh.

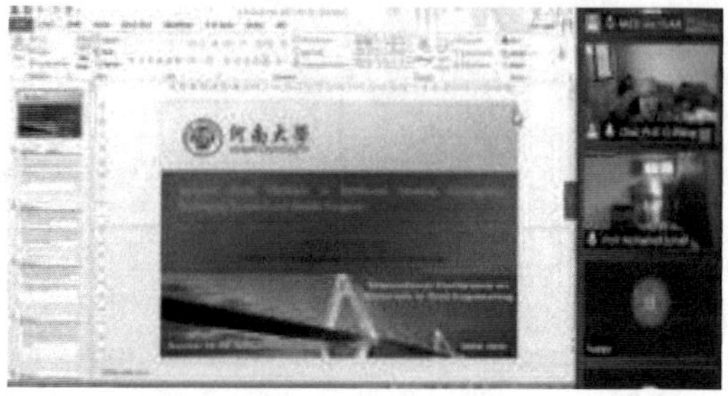

Figure 4. Keynote presentation from Prof. Mohamed Abdel Kader El Gelany Ismail.

The third keynote speaker, Prof. Faiz Uddin Ahmed Shaikh, from Curtin University, Australia, gave a speech on" Effect of nano calcium carbonate on compressive strength development and durability properties of high volume slag and slag-fly ash blended concretes"(**Figure 3**).

The fourth and also the last keynote speaker, Prof. Mohamed Abdel Kader El Gelany Ismail, from Miami College of Henan University, talked about" Recycled waste materials in reinforced concrete construction: developing research and wealth prospects"(**Figure 4**).

After keynote presentations, there are 8 oral presentations(**Figure 5**) and 7 poster presentations(**Figure 6**). Oral speakers and poster presenters shared their latest research results with the audience, the audience actively and enthusiastically participated in by asking questions and online voting.

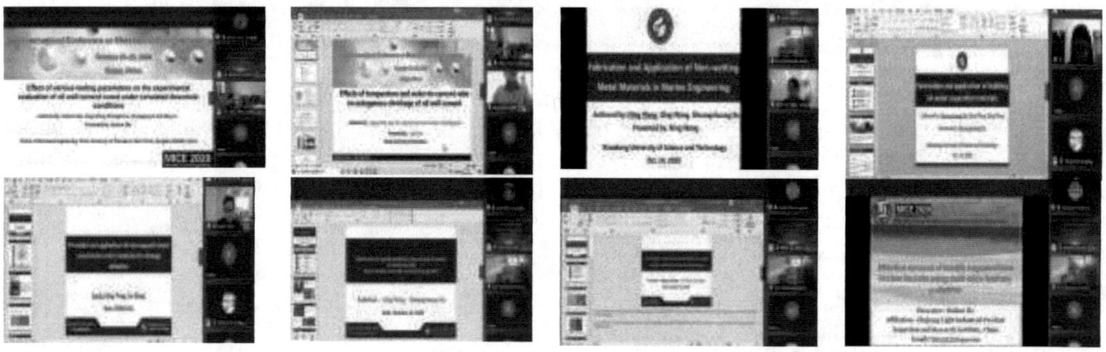

Figure 5. 8 Oral presentations of MICE2020.

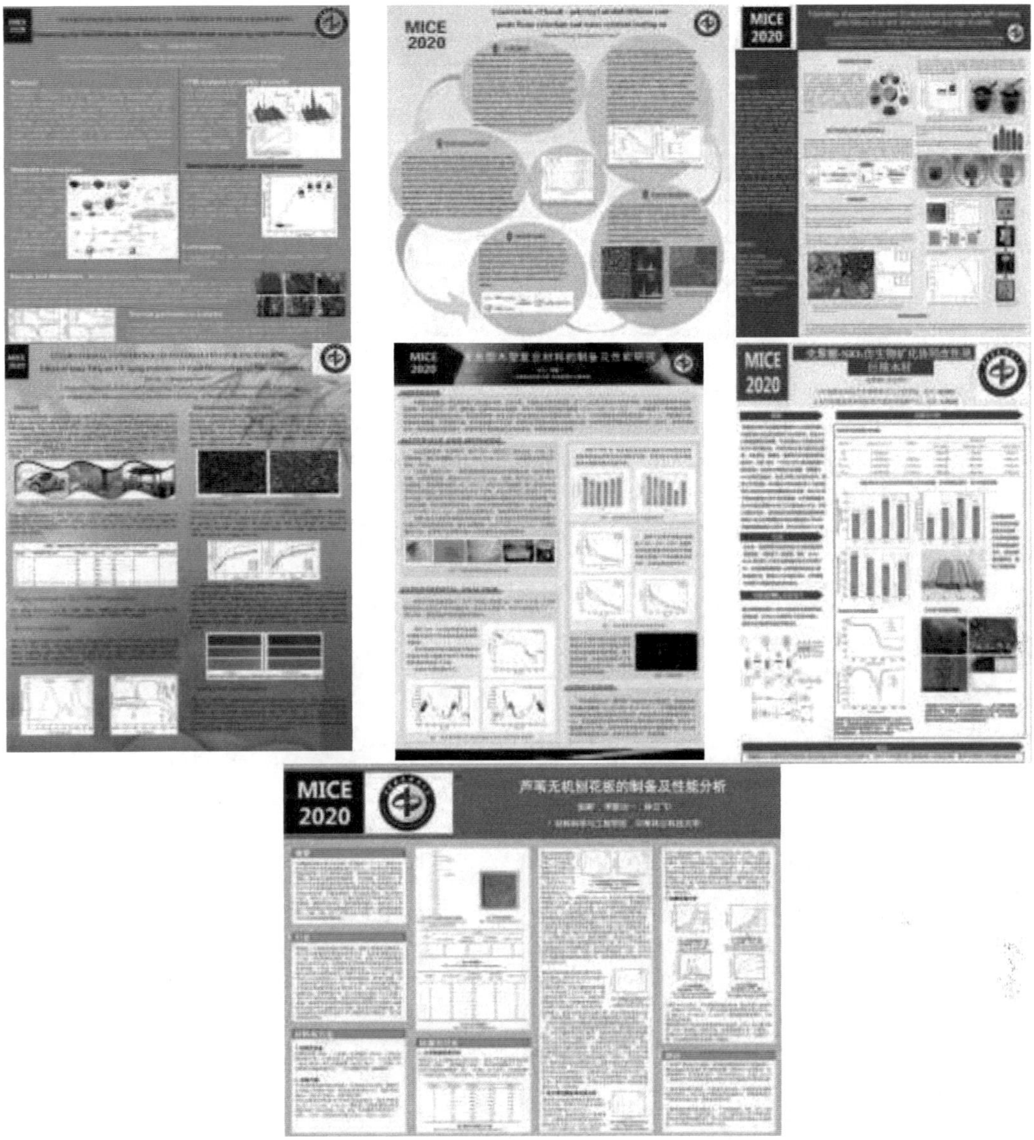

Figure 6. 7 Poster presentations of MICE2020

We would like to express our deepest feelings of respect and gratitude to the host, outstanding keynote speakers, oral speakers as well as all the audiences.MICE2020 online conference was held successfully because of their great supports and help.

IOP Conf. Series: Materials Science and Engineering 1028 (2021) 011002 doi:10.1088/1757-899X/1028/1/011002

Peer review declaration

All papers published in this volume of *IOP Conference Series: Materials Science and Engineering* have been peer reviewed through processes administered by the Editors. Reviews were conducted by expert referees to the professional and scientific standards expected of a proceedings journal published by IOP Publishing.

- Type of peer review: Double-blind

Author's information were removed from the papers before be forwarded to the reviewers. And the authors only got the comments of papers without reviewers' information.

- Conference submission management system:
  http://papersub.icmice.org/

- Number of submissions received:
  41

- Number of submissions sent for review:
  33

- Number of submissions accepted:
  15

- Acceptance Rate (Number of Submissions Accepted / Number of Submissions Received X 100):
  Approximately 37%

- Average number of reviews per paper:
  3

- Total number of reviewers involved:
  89

- Any additional info on review process:
  After preview by editorial assistant, 33 papers were sent for peer reviews, 89 reviewers involved in the process, and review process for each paper took around 1 month. The decision to accept or reject was based on the reviewers' comments. If differences of opinion occurred between reviewers, the editors weighed all comments and arrived at a balanced decision based on all comments, or a second round of peer review was initiated.

- Contact person for queries:
  Ms.Yuan, secretary@icmice.org

IOP Conf. Series: Materials Science and Engineering 1028 (2021) 012001 doi:10.1088/1757-899X/1028/1/012001

Effect of acid on the early strength of aluminum sulfate series alkali-free accelerator

Liya Wang[1,2,*], Guisheng Cai[1,2], Yukun Gong[1,2] , Jun Liu[1,2] and Tongwei Lu[1,2]

[1] Shandong academy of building research co. ltd;
[2] Shandong jianke building materials co. ltd.

[*]Email: wly.sdu@163.com

Abstract. This article introduces the effect of acid on the setting time, early strength and stability of aluminum sulfate alkali-free accelerator. The inorganic acids used are phosphoric acid, hydrofluoric acid and fluorosilicic acid, and the organic acids are oxalic acid and ethylenediaminetetraacetic acid. As a result, it was found that increasing the amount of acid can increase the stability of the accelerator. The product of the anion in acid and calcium ion in cement is easy to form. The product damages the network structure of the cement paste, and the early strength is low.

1. Introduction

In recent years, China's tunnels, underground facilities and other infrastructure have been vigorously promoted. When constructing a tunnel, in order to ensure that no collapse occurs, strong support must be provided. Sprayed concrete can achieve good molding effect without the support of formwork. The time required for concrete hardening is relatively short, which can greatly shorten the construction period, and has a good effect on the reinforcement of the tunnel surrounding rock. The so-called sprayed concrete is actually mixed with additives, cement, etc. with the help of air pressure, and then sprayed to the working surface. In this way, the mixture can quickly harden and solidify, thereby enhancing the supporting force and strength.

In the study of sprayed concrete, the accelerator is often in a key position. The usage and market share of liquid accelerators have developed rapidly. Among them, the proportion of liquid alkali-free quick setting agent is increasing. At present, there are still some problems with the alkali-free liquid accelerator on the market. For example, the price is high, the stable storage period is short, the early strength development is slow, and the cement adaptability range is small. Alkali-free liquid accelerator has greater difficulty in product development [1].

In the aluminum sulfate alkali-free accelerator system, to suppress the hydrolysis of aluminum ions into aluminum hydroxide precipitation, the pH of the solution needs to be controlled below 4.0. When the pH of the system is greater than 4, precipitation inevitably separates out. Under normal storage conditions after production, it is prone to crystallization and delamination, that is, the problem of short storage period [2]. To solve this problem, the researchers will add acid to the liquid alkali-free accelerator system. By adjusting the pH value of the system, the effect of inhibiting the hydrolysis of aluminum ions is achieved. Commonly used acids are divided into inorganic acid and organic acid. Inorganic acids include phosphoric acid, hydrofluoric acid, and fluorosilicic acid [3-8], and organic acids include oxalic acid, lactic acid, methacrylic acid, glycolic acid, and ethylenediaminetetraacetic acid (EDTA) [9-11].

The experiment found that after using acid to solve the problem of short storage period of aluminum sulfate alkali-free quick setting agent, the problem of early strength (1d strength) not meeting the standard will occur. In this paper, three inorganic acids, phosphoric acid, hydrofluoric acid and fluorosilicic acid are selected. Most of the organic acids used are organic carboxylic acids. In this paper, oxalic acid and EDTA are selected. This article explores the effect of acid dosage on the early strength of aluminum sulfate alkali-free accelerator.

2. Materials and methods

2.1. Raw materials

PO42.5 cement is produced by Shandong Shanshui Cement Group Co. Ltd. Standard sand is Xiamen ISO standard sand. Aluminum sulfate, magnesium fluorosilicate, triethanolamine, phosphoric acid, hydrofluoric acid, fluorosilicic acid, oxalic acid and EDTA are industrial grade. Water was deionized water.

2.2. Preparation of alkali-free accelerator

40% water and 5% magnesium fluorosilicate were added to the three-necked flask, after heating to 60°C, 50% aluminum sulfate was added in portions. After 2 hours of reaction, 5% triethanolamine and the acid studied in this article were added in sequence. Stir until the solution is homogeneous. After cooling to room temperature, an alkali-free quick-setting agent is obtained, which is marked as NC-S. The amount of acid added is a percentage of the total.

2.3. Performance test methods

The performance of alkali-free accelerator is tested according to GB / T 35159—2017 "Accelerator for Shotcrete". GB/T 35159-2017 stipulates: water/cement ratio is 0.35. Initial setting time ⩽5 min, final setting time ⩽12min; 1 d compressive strength ⩾7.0 MPa, 28 d compressive strength ratio ⩾90%, 90 d The retention rate of compressive strength is ⩾100%; the stability at 28 days is ⩽5 ml. In this experiment, the amount of accelerator is 7% of the cement mass. After the synthesis of the sample is completed, it is placed in a transparent sample bottle, sealed and stored in an environment of (20±2)° C, and the changes in the clarity and stratification of the sample are regularly observed, and the stable period is recorded. When the thickness of the supernatant reaches 5 ml after the liquid is layered, it is defined as the unstable state, and the sample is from the end of the synthesis to the unstable state as the stable period (in days).

3. Results and discussion

3.1. The effect of phosphoric acid on the performance of alkali-free accelerator.

The amount of phosphoric acid was changed to study the changes in product setting time, early strength and stability. The results are shown in Table 1.

It can be seen from Table 1 that as the amount of phosphoric acid increases, the initial setting time and final setting time of the alkali-free quick-setting agent both decrease first and then increase. The shortest time is reached at 4% dosage. The sample gradually became clear from turbidity. When the amount of phosphoric acid is more than 4%, the sample has become transparent, and the stability period is also extended. This is because phosphoric acid is a medium strong acid. Increasing the amount of phosphoric acid increases the concentration of hydrogen ions in the system, which can effectively inhibit the hydrolysis process of aluminum ions.

Table 1. The effect of phosphoric acid on the performance of alkali-free accelerator.

Dosage /%	Setting time / (min:s)		1d Compressive strength /MPa	Product properties	pH	Stability period /d
	initial	final				
0	5:40	11:50	8.9	yellow-green, turbid, sticky	5.5	3
1	5:20	10:40	7.9	yellow-green, turbid, sticky	4.3	10
2	4:40	8:30	7.2	yellow-green, cloudy, sticky	4.0	18
3	4:20	7:40	6.8	yellow-green, turbid, thick	3.9	23
4	3:30	6:50	5.7	yellow-green, transparent, sticky	3.5	30
5	4:10	8:20	4.5	yellow-green, transparent, sticky	3.3	30

However, as the amount of phosphoric acid increases, the early strength decreases. The retardation effect of phosphoric acid itself on cement is not great. But phosphoric acid will form phosphate when added to the alkali-free accelerator system. This phosphate can ionize the charged ions in the aqueous solution, causing displacement and condensation. During the setting and hardening process of cement, phosphate produces an insoluble film. The film layer will affect the precipitation and nucleation of Ca $(OH)_2$, C-S-H and the formation process of C-A-S-H, and then delay the setting and hardening of cement to produce retardation [12].

3.2. The effect of hydrofluoric acid on the performance of alkali-free accelerator.

Change the content of hydrofluoric acid to study the changes of product setting time, early strength and stability. The results are shown in Table 2.

It can be seen from Table 2 that as the amount of hydrofluoric acid increases, the initial setting time and final setting time of the alkali-free quick-setting agent continue to decrease. The prepared samples are light green, transparent, low viscosity and easy to flow. The period of stability has also been extended. When the content of hydrofluoric acid is 2%, the stable period reaches 60 days. This is because F^- will complex with Al^{3+}, which greatly increases the solubility of Al^{3+} and increases its stability.

As the amount of hydrofluoric acid increased, the early strength decreased slightly. This is because F- will form CaF_2 or F^- with Ca^{2+}, which will depolymerize the C-S-H gel and damage the network structure of the cement paste, and the early strength is low [4,5].

Table 2. The effect of hydrofluoric acid on the performance of alkali-free accelerator.

Dosage /%	Setting time / (min:s)		1d Compressive strength /MPa	Product properties	pH	Stability period /d
	initial	final				
0	5:40	11:50	8.9	yellow-green, turbid, sticky	5.5	3
0.5	4:40	7:50	7.3	light green, transparent and easy to flow	4.2	15
1	3:50	7:30	7.2	light green, transparent and easy to flow	3.6	30
1.5	3:00	5:00	6.5	light green, transparent and easy to flow	3.0	40
2	2:30	4:50	6.3	light green, transparent and easy to flow	2.6	60

Table 3. Effect of fluorosilicic acid on the performance of alkali-free accelerator.

Dosage /%	Setting time / （min:s）		1d Compressive strength /MPa	Product properties	pH	Stability period /d
	initial	final				
0	5:40	11:50	8.9	yellow-green, turbid, sticky	5.5	3
1	4:20	7:50	7.0	Light green, transparent, sticky	5.0	10
2	4:50	7:30	7.3	Light green, transparent, sticky	4.6	15
3	5:00	8:00	7.8	Yellow-green, turbid, sticky	4.5	20
4	5:30	8:50	4.3	Yellow-green, turbid, sticky	4.2	16

3.3. The effect of fluorosilicic acid on the performance of alkali-free accelerator.

Change the content of fluorosilicic acid to study the changes of product setting time, early strength and stability. The results are shown in Table 3.

It can be seen from Table 3 that with the increase of the amount of fluorosilicic acid, the initial setting time and final setting time of the alkali-free quick-setting agent are continuously increasing. The prepared sample is light green and transparent. After the dosage is increased to 3%, the color changes to yellow-green, and the sample appears small granular crystals. This is because the addition of fluorosilicate can first complex with Al3 +, forming a large number of network structures in the solution, increasing the solubility of Al3 +. As the viscosity of the sample increases, the stability period also increases. However, after the amount of fluorosilicic acid is greater than 3%, granular crystals of aluminum fluorosilicate or magnesium fluorosilicate appear in the solution, and the stability of the system decreases.

With the increase of the amount of fluorosilicic acid, the early strength is slightly enhanced. After the amount is increased to 4%, the strength decreases. This is because the fluorosilicate will form calcium fluorosilicate with Ca^{2+}, which will depolymerize the C-S-H gel and damage the network structure of the cement paste, and the early strength is low [6].

3.4. The effect of oxalic acid on the performance of alkali-free accelerator.

The amount of oxalic acid was changed to study the changes in product setting time, early strength and stability. The results are shown in Table 4.

It can be seen from Table 4 that with the increase in the amount of oxalic acid, the initial setting time and final setting time of the alkali-free quick setting agent are continuously increasing. The prepared sample is light green and transparent first. After the dosage is increased to 1%, the viscosity of the system drops. This is because oxalate can chelate with Al^{3+}, forming a large number of network structures in solution. The solubility of Al^{3+} is increased, and the stability period is also extended.

Table 4. The effect of oxalic acid on the performance of alkali-free accelerator.

Dosage /%	Setting time / （min:s）		1d Compressive strength /MPa	Product properties	pH	Stability period /d
	initial	final				
0	5:40	11:50	8.9	yellow-green, turbid, sticky	5.5	3
1	4:25	10:20	7.0	Light green, transparent, sticky	4.2	14
2	4:50	11:30	6.5	Light green, transparent and easy to flow	3.6	20
3	5:00	12:00	4.8	Light green, transparent and easy to flow	3.5	21

As the amount of oxalic acid increased, the early strength continued to decline. This is because oxalate will form calcium oxalate with Ca^{2+}, which will depolymerize the C-S-H gel and damage the network structure of the cement paste, and the early strength is low [13].

3.5. The effect of EDTA on the performance of alkali-free accelerator.

Change the dosage of EDTA to study the changes of product setting time, early strength and stability. The results are shown in Table 5.

Table 5. Effect of EDTA on the performance of alkali-free accelerator.

Dosage /%	Setting time / (min:s)		1d Compressive strength /MPa	Product properties	pH	Stability period /d
	initial	final				
0	5:40	11:50	8.9	yellow-green, turbid, sticky	5.5	3
1	3:30	7:30	7.5	Light yellow, transparent and easy to flow	5.6	20
2	4:10	8:10	7.3	Yellow, transparent, easy to flow	5.4	30
3	4:15	8:30	6.7	Yellow, transparent, easy to flow	4.8	40
4	4:30	9:00	5.1	Yellow, transparent, easy to flow	4.6	40

It can be seen from Table 5 that with the increase in the amount of EDTA, the initial setting time and final setting time of the alkali-free quick-setting agent continue to increase. The prepared sample is light green, transparent and easy to flow. When the dosage is increased to 2%, the color of the system becomes yellow. This is because EDTA is a chelating agent that can chelate with Al^{3+}, forming a large number of network structures in the solution, increasing the solubility of Al^{3+} and prolonging the stability period.

As the amount of EDTA increased, the early strength continued to decline. This is because EDTA also chelate with Ca^{2+}, which will depolymerize the C-S-H gel and damage the network structure of the cement paste, and the early strength is low [9].

4. Discussion and conclusion

Based on the above experimental results, it can be seen that increasing the amount of acid and reducing the pH of the system is an important way to improve the stability of alkali-free accelerator. Lowering the pH value of the solution can inhibit the hydrolysis of aluminum ions into aluminum hydroxide precipitation and improve the stability of the accelerator.

Adding H^+ to the system will inevitably introduce some anions. These anions will affect the solubility of aluminum ions and at the same time will affect the early strength of cement. For inorganic acids, the anion complexes with aluminum ions, and the aluminum salt formed can increase the solubility of aluminum ions. For organic acids, the anion is generally a carboxylate group, which chelate with aluminum ions, forming a large number of network structures in solution, increasing the solubility of aluminum ions. In terms of strength, the product formed by these anions and calcium ions destroys the network structure of the cement paste, and the early strength is low.

References

[1] Zeng Luping, Wang Wei, Zhang Xiaolei, Zhao Shuang, Qiao Min 2019.Research status of liquid alkali-free accelerator and research on cement adaptability mechanism *Building materials development orientation* **17 (16)** 11-14

[2] Xue Bin, Xue Shaolong, Yin Lingyun, Sheng Yanping 2014.Performance and mechanism of liquid alkali-free accelerator *Concrete* **(02)** 82-84 + 87

[3] Long Fuguo, Ren Junpeng, Zhao Jun, Wang Yu, Wu Kun, Cai Rongjin 2019.Preparation and performance study of a new type of liquid alkali-free quick setting agent *Modern salt*

chemical industry **46 (06)** 5-7 + 20
[4] Guo Jing 2019 Application Research of Non-alkali Liquid Accelerator for Shotcrete Used in Tunnel. Changan University
[5] Yu Shaoqiu, Feng Enyao, Li Conghai, Huang Ling, Li Shufan 2019.Preparation and mechanism study of low-sulfur alkali-free accelerators *Building materials world* **40 (04)** 1-5
[6] Xu Zhongzhou, Zhu Jianing, Fan Ruibo, Gao Wenfu.Preparation and properties of high adaptability liquid low-alkali accelerator 2019 *New building materials* **46 (11)** 27-29 + 41
[7] Zhang Yong, Zhang Xiaolei, Yang Xiaoqi, Liu Yongsheng, Wang Wei, Qiao Min, Ran Qianping 2019 Effect of anti-settling agent on the performance of alkali-free accelerator in fluorosilicate system *New building materials* **46 (07)** 126 -127 + 149
[8] Gan Jiezhong, Wang Ling, Tian Pei, Zhao Xia, Chen Shuanfa, Guan Bowen 2014 The effect of acid on the stability of aluminum sulfate series liquid alkali-free accelerators *Journal of Wuhan university of technology* **36 (02)** 38-43
[9] Deng Lei, Jiang Yu, Shen Jianrong, Diao Lizhu 2018 Development of a new type of alkali-free accelerator *China building materials science and technology* **27 (03)** 27-28
[10] Deng Lei, Jiang Yu, Luo Xiaofeng, Shen Jianrong, Zhang Bingwang 2018 Effect and application of water-soluble organic matter on the performance of liquid alkali-free accelerator *Commercial concrete* **(06)** 53-56
[11] Wang Longfei, Li Qianqian, Dong Shuqiang, Zhang Qi, Yu Jingrui 2017 Research on preparation and performance of alkali-free accelerator for shotcrete *Kejifeng* **(10)** 103-104
[12] Lin Yankun, Zhang Dechen, Xu Jianzhao 2018 The mechanism and performance of phosphate / phosphine and its salt retarder *Commercial concrete* **(08)** 25-28
[13] Göril, Möschner 2009 Influence of citric acid on the hydration of Portland cement *Cement & concrete research*

Effects of temperature and water-to-cement ratio on autogenous shrinkage of oil well cement

Xueyu Pang[1,2,*], Lijun Sun[2], Socorria Dos Santos Araujo[2] and Yuhuan Bu[1,2]

[1] Key laboratory of unconventional oil & gas development (China university of petroleum (east China)), ministry of education, Qingdao 266580, P. R. China;
[2] School of petroleum engineering, China university of petroleum (east China), Qingdao 266580, P. R. China.

*Email: x.pang@upc.edu.cn

Abstract. The chemical shrinkage and corrugated tube test methods were used to study the shrinkage behaviour of Class G oil well cement with water-to-cement (w/c) ratios of 0.3 and 0.5, cured at temperatures from 15 to 60°C. The experimental results show that the typical linear shrinkage evolution curve of a cement slurry obtained by corrugated tube method have an obvious inflection point, beyond which the linear shrinkage increase is close to zero. The time at which the inflection point is observed decreases significantly with increasing curing temperature, but varies very little with w/c ratio. The inflection time also shows approximately linear correlation with the setting time of the cement (The two are almost equal for w/c =0.3). The degree of hydration corresponding with the initial and final setting time were estimated based on chemical shrinkage test results and the relationships between estimated volumetric bulk shrinkage and chemical shrinkage were analyzed.

1. Introduction

With the slow depletion of oil and gas resources in conventional reservoirs, oil and gas exploration are forced to develop in the direction of complex formations, which presents bigger challenges for the stability and safety of oil exploration. Cementing is one of the most important links in oil and gas well engineering. Its main function is to achieve zonal isolation between strata and establish efficient oil flow channels [1]. The quality of cementing determines the service life of oil and gas wells, and it could even determine the success or failure of an entire oil well [2]. Annular channeling is one of the main problems in the field of cementing engineering. The volume shrinkage of oil well cement will directly lead to the deterioration of the bonding strength of the first and second interfaces of the cement sheath [3-5]. This is the main cause of micro-cracks and fluid channeling in the annulus of the wellbore. Therefore, it is of great importance to develop an experimental method that can accurately quantify and characterize the volumetric shrinkage performance of oil well cement.

The hydration of cement is a complex chemical process that involves the reduction in volume when reactants (dry cement and water) are converted to resultants (hydration products). The total volume reduction is often referred to as the chemical shrinkage of cement, which can be reliably measured by a number of different test methods [6-8]. However, in engineering applications, it is the external bulk volume shrinkage that is detrimental to the structural integrity of cement-based materials. The bulk volume shrinkage is a lot more difficult to measure accurately compared to chemical shrinkage. API has reported four different methods to measure the bulk volume shrinkage of oil well cement, including the annular ring-mold test, the membrane test, the cylindrical-sleeve test and the cement hydration analyzer test [9]. The first two methods have later been included in an API standard [10]. However, according to the review conducted by Reddy et al. [2], the agreements of test results obtained by these different methods are very poor. While the results of the annular ring-mold test appear to agree reasonably well with the ASTM method (ASTM C-151), the membrane test tends give significantly higher results, approaching the total chemical shrinkage values. Lura and Jensen had later found that the main source of error of the membrane test come from water uptake through the membrane [11-13].

Experimental measurement of the bulk volume shrinkage of cement-based materials is further complicated by the fact that cement slurry/paste experiences a phase transition from liquid to solid during the setting period. When the slurry is in a liquid state, all chemical shrinkage will be converted to bulk shrinkage (this is also known as the plastic shrinkage); With the development of a solid framework during setting, external bulk shrinkage will be significantly less than chemical shrinkage, and the majority of total volume reduction occurs inside the solid framework (often called internal shrinkage). API ring-mold test and ASTM C-151 are suitable methods for bulk shrinkage measurement of cement. However, with these methods, measurements can only be obtained after the cement had hardened and developed compressive strength. A standard corrugated tube test method is now available through ASTM [14], which allows the bulk shrinkage of cement to be continuously measured from the time of casting [15-17]. The corrugated tube provides minimum restraints to the dimension change of the specimen and prevent moisture exchange with the environment, avoiding most artifacts associated with other methods. Because the specimen is in sealed condition, the results obtained by this method is often called autogenous shrinkage, as it is primarily caused by self-desiccation inside the material.

Shrinkage is a complicated phenomenon in the cement hydration process, which is affected by many factors such as cement type, fineness, mineral composition, water-to-cement ratio (w/c), and curing conditions (temperature and pressure) [18-19]. This article focuses on studying the influences of w/c and curing temperature on the shrinkage behavior of oil well cement using chemical shrinkage and corrugated tube test methods. Comparisons between chemical shrinkage and bulk shrinkage can provide some interesting insights regarding the mechanism of cement shrinkage. Since autogenous shrinkage is defined as the bulk shrinkage of a sealed cement sample from the time of final setting, Vicat needle tests [20] were employed to determine the setting time of cement with different w/c ratios and cured at different temperatures.

2. Materials and methods

2.1. Materials and slurry preparation

The cement used in this study is an API Class G oil well cement, which is provided by Akesu cement plant, Xinjiang province, China. The compound composition of the cement as determined by quantitative XRD analysis is given in Table 1.

Table 1. Estimated main compound compositions (by mass percentage) of the cement.

Compound	C_3S	C_2S	C_3A	C_4AF	Gypsum
Quantity/%	46.0	28.0	0.0	19.7	6.4

Table 2. Cement slurry formulation (% by weight of cement) and density of raw materials.

Raw materials	Cement	Water	Dispersant	Defoamer	Diutan gum
Density/(kg/m^3)	3.24×10^3	1.0×10^3	1.1×10^3	0.96×10^3	1.0×10^3
Formula 1	100	50	1	0.5	0.02
Formula 2	100	30	1.5	0.5	0

Slurries were prepared according to standard API procedures [21] with water-to-cement (w/c) ratios of 0.3 and 0.5, respectively. A dispersant was used to tune the rheology of the slurries while a defoamer was used to remove the entrained air. A small amount of diutan gum was added to improve the stability of the slurry at higher w/c. The dispersant (BCD-210L) and defoamer (G603) used in this study were both provided by Tianjin PetroChina Boxing Technology Co., Ltd. The diutan gum was provided by CP Kelco (Shandong) Biological Co., Ltd. Detailed slurry compositions are shown in Table 2.

2.2. *Vicat needle test*

Vicat needle test was used to measure the initial and final setting time of the cement according to ASTM C191-18 [20]. The initial time of setting is the time elapsed between the initial contact of cement and water and the time when the penetration is measured or calculated to be 25 mm. The initial time of setting can be calculated by formula (1):

$$\sqrt{\frac{H-E}{C-D}} \times (C-25) + E \tag{1}$$

where, H (min) is the time for the first penetration less than 25 mm; E (min) is the time for last penetration greater than 25 mm; C (mm) is the penetration reading at time E; D (mm)is the penetration reading at time H. The final time of setting is the time elapsed between initial contact of cement and water and the time when the needle does not leave a complete circular impression in the paste surface.

2.3. *Chemical shrinkage test*

During this study, a self-designed experimental device was used to measure the chemical shrinkage of oil well cement at different curing temperatures. The device utilizes a syringe pump to continuously monitor the amount of water that is injected to a curing chamber filled with cement slurry. Such experimental test method is the same as those used in previous studies for oil well cement chemical shrinkage measurements and details of the experimental setup can be found in those references [7-8]. The inside of the curing chamber used here is shaped as a tapered cylinder with an average diameter of 55mm and a height of 250mm, rendering a total volume of about 600 mL. The curing chamber was pre-set to the target temperature to allow temperature stabilization before each test. A constant curing pressure of 5 MPa was applied during each test. Based on the results of previous studies, the influence of such a low pressure on cement hydration rate should be very minimal [7-8, 22].

2.4. *Corrugated tube shrinking test*

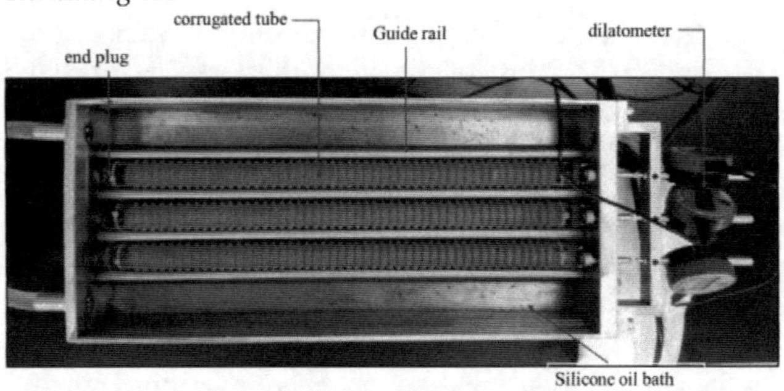

Figure 1. Experimental set-up of the corrugated tube autogenous shrinkage test.

The corrugated tube shrinkage test device used in this article was designed according to ASTM C 1698-09 [14]. A picture of the experimental set-up is shown in Figure 1. The device consisted of three measurement channels to produce replicate results. The specific experimental steps are as follows: first, precondition the oil bath to the target temperature; secondly, pour the prepared cement slurry into corrugated tubes and seal the ends with plugs; finally, place the slurry-filled tube on the guide rail in the oil bath and continuously measure the length change of the tube using the dilatometer.

3. Test results and discussion

3.1. Settlement stability test results

It is necessary to determine the settlement stability of the cement slurry before the test to avoid test artifact caused by instability of the cement slurry. According to API standard [21], cylindrical cement samples with a diameter of 25 mm and a height of 150 mm were prepared and placed in a constant-temperature water bath for 24 hours. The cured samples were cut into four equal sections, with their densities determined according to the Archimedes' principle. No density difference was observed among different sections for each slurry formulation, indicating excellent slurry stability. The set cement density for formula 1 and formula 2 were 1.87×10^3 and $2.12 \times 10^3 kg/m^3$, respectively.

3.2. Vicat needle test results

The test method described in Section 2.2 was used to test the setting time of two cement slurries with different w/c ratios under different temperatures using the Vicat needle method. The specific test results are shown in Table 3.

Table 3. Test results of Vicat needle setting test.

Formula #	Formula 1(w/c=0.5)			Formula 2(w/c=0.3)		
Temperature/°C	30	45	60	30	45	60
Initial setting time/h	7.9	6.1	5.0	7.7	4.7	4.5
Final setting time/h	9.8	8.3	5.8	9.9	6.1	5.4

It can be seen from the test results that the setting time of two cement slurries decreases with the increase of curing temperature. Specifically, the initial setting time of the cement slurry was about 7.8 hours at 30°C and about 4.5-5 hours at 60°C; The final setting time was about 10 hours at 30°C and 5.4-5.8 hours at 60°C. The reduction in setting time with increasing temperature is mainly due to the fact that cement hydrates faster at higher curing temperatures. Comparing the test results of Formula 1 and Formula 2, it can be clearly observed that under the same curing temperature, the setting time of the cement slurry with lower w/c is generally shorter than that with higher w/c. This is because the cement slurry with a lower w/c has a higher solid volume fraction, and requires a relatively low degree of hydration to achieve setting. It is noted that the effect of w/c on the setting time is much smaller than the effect of curing temperature.

3.3. Corrugated tube shrinking test results

The linear shrinking test of formula 1 and formula 2 were completed under four different temperatures, namely 15°C, 30°C, 45°C and 60°C. Most of the time, the three measurement channels of the test device yielded at least 2 valid results per data set. It generally took about 20 minutes from the beginning of mixing cement slurry to the start of data collection. For consistency, the zero time of these experiments were set at the starting point of data recording.

Figure 2 and Figure 3 present the test results of the linear shrinkage evolution of the two cement slurries at 15°C. The test results show that both slurries show similar shrinkage behavior: the linear shrinkage of cement increases relatively rapidly during the early stage of cement hydration and then slows down rapidly after the cement sets. Correspondingly, the rate of linear shrinkage starts at a relatively high value and then gradually decrease to near zero over time. There is an obvious inflection point on the linear shrinkage evolution curve. The total linear shrinkage before the inflection point (plastic shrinkage) is in the range from 0.65% to 0.8% while the shrinkage increase after the inflection point is close to zero. For the purpose of consistency, the inflection time is defined by when the amount of shrinkage measured over a 5 minutes period first becomes less than 1 μm, which can be easily identified from the rate of shrinkage curve. At 15°C, the average inflection time is found to be 16.4 h for formula 1 (w/c =0.5) and 13.7 h for formula 2 (w/c =0.3). The inflection point in the linear shrinkage curve is likely to be caused by the setting of the cement, which provides a solid skeleton to the cement mixture to resist external shrinkage.

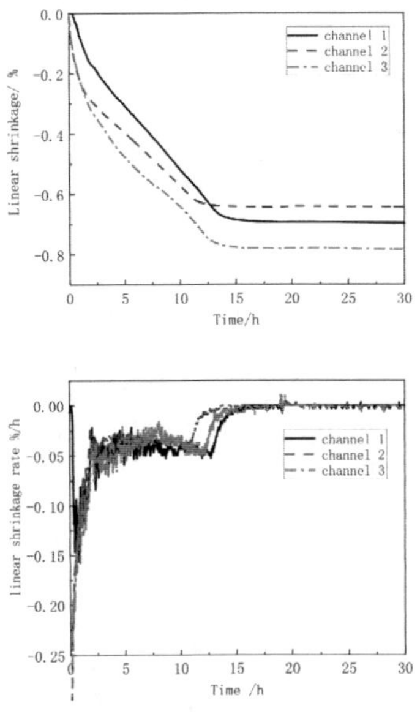

Figure 2. Linear shrinkage test results at 15°C (w/c=0.3).

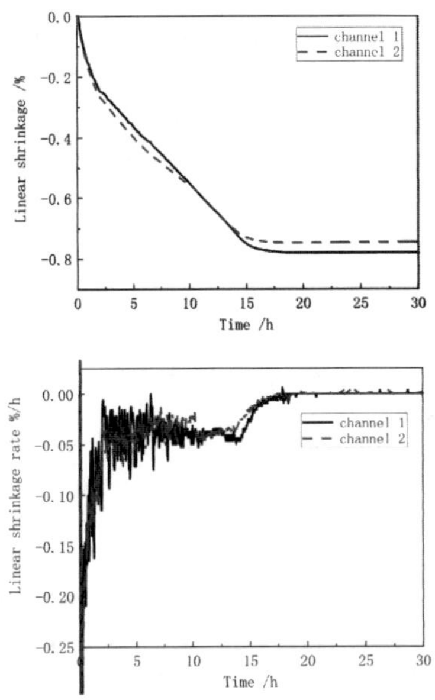

Figure 3. Linear shrinkage test results at 15°C (w/c=0.5).

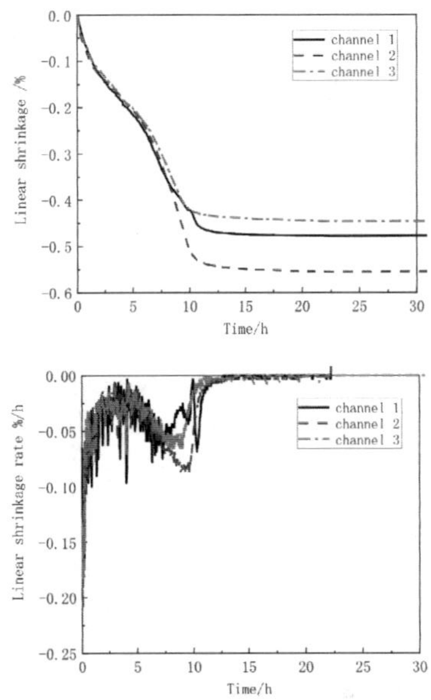

Figure 4. Linear shrinkage test results at 30°C (w/c=0.3).

The time evolution curves of the linear shrinkage of the two cement slurries with different w/c at 30°C are shown in Figure 4 and Figure 5. The overall shrinkage evolution profile is similar to the trend observed at 15°C. Specifically, the linear shrinkage evolution curves showed obvious inflection points after about 9-11 hours hydration, earlier than those observed at 15°C due to faster setting. The inflection time agrees reasonably well with the Vicat needle final setting time measured at the same curing temperature (See Table 3). The amount of linear shrinkage measured before the inflection point (plastic shrinkage) is in the range between 0.4% and 0.55%, significantly smaller than that observed at 15°C. The shrinkage increase after the inflection point is approximately zero.

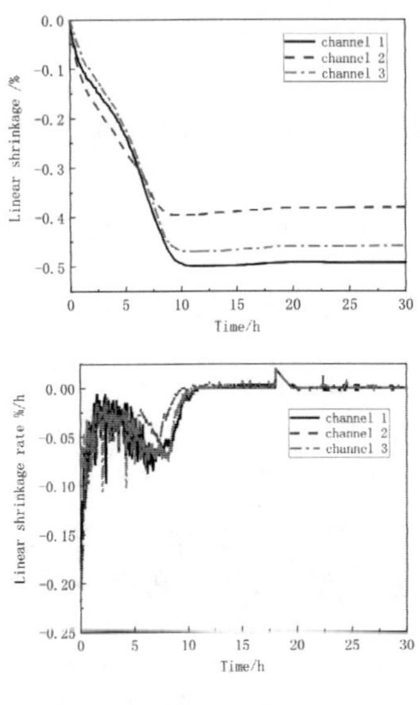

Figure 5. Linear shrinkage test results at 30°C (w/c=0.5).

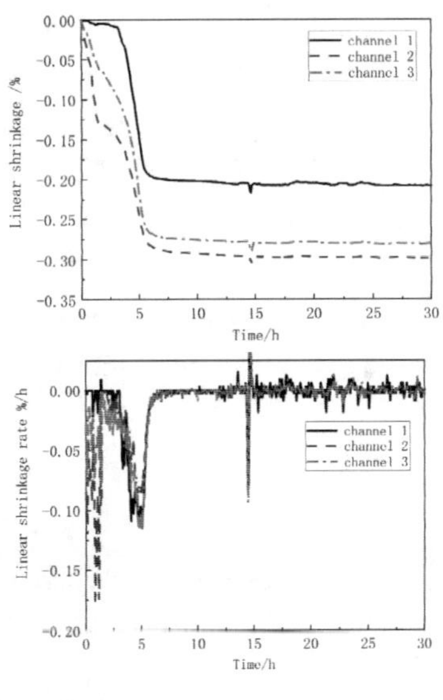

Figure 6. Linear shrinkage test results at 45°C (w/c=0.3).

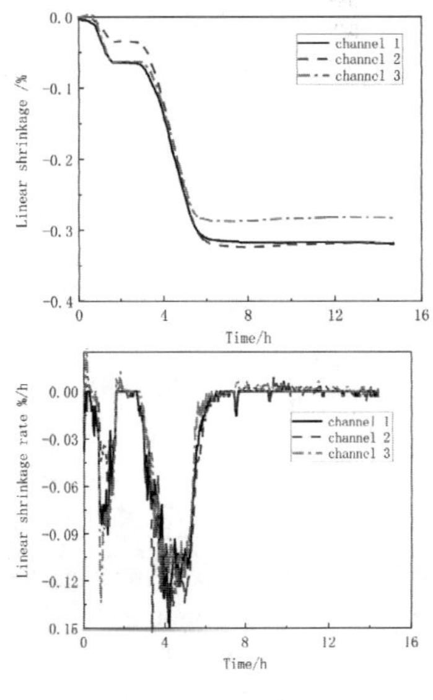

Figure 7. Linear shrinkage test results at 45°C (w/c=0.5).

The time evolution curves of the linear shrinkage of the two cement slurries with different w/c at 45°C are shown in Figure 6 and Figure 7. Although some inconsistencies are observed between three different measurement channels for the low w/c slurry, the overall shrinkage evolution profiles are similar to the above test results. The total linear shrinkage values before the inflection points were about 0.2%-0.3%, smaller than those measured at lower temperatures. The average time at which the inflection points appeared is 5.9 h for formula 1 (w/c=0.5) and 5.8 h for formula 2 (w/c=0.3), respectively.

The shrinkage evolution behavior of the two cement slurries at 60°C are also similar to previous test results at lower temperatures, but the total linear shrinkage before the inflection point is further reduced to about 0.1%-0.14%. In order to get more direct comparisons of cement linear shrinkage behavior at different temperatures, the results of each data set were averaged and presented in Figure 8. It is obvious that the total linear shrinkage during the early stage before the inflection point decreases significantly with increasing curing temperature. According to previous studies, the linear shrinkage of cement during the early stage can be strongly influenced by the relative rigidity of the corrugated tube [17]. Therefore, the variation of early test results with temperature is likely to be associated with the rigidity change of the cement mixture and the corrugated tube at different temperatures. Figure 8 also shows that when the linear shrinkage is zeroed at the final setting time of cement based on Vicat needle test, the total amount of autogenous shrinkage at all curing temperatures is near zero after final setting with the exception of formula 2 (w/c=0.3) at 30°C. These measured shrinkage values are lower compared to ordinary Portland cement used in the construction industry [23], which may be explained by the differences in cement composition, considering that the Class G oil well cement used here contains no C_3A.

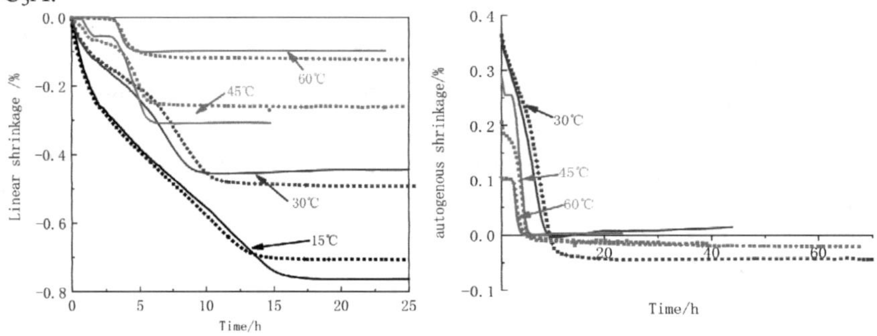

Figure 8. Summary of average linear shrinkage and autogenous shrinkage at different temperatures. (dotted lines: w/c=0.3, solid lines: w/c=0.5)

Table 4. Inflection time (/h) on the shrinkage evolution curves at different test conditions.

w/c	0.3				0.5			
Temperature	15°C	30°C	45°C	60°C	15°C	30°C	45°C	60°C
Channel 1	12.63	11.05	5.79	5.24	16.13	9.59	5.98	4.8
Channel 2	13.65	10.95	5.83	5.38	16.67	8.63	6.08	
Channel 3	14.94	10.13	5.84	5.73		9.36	5.61	
Average	13.74	10.7	5.8	5.45	16.4	9.2	5.9	4.8
COV	8.4%	4.7%	0.5%	4.6%	N/A	5.5%	4.2%	N/A

Table 4 summarizes the observed inflection time on the shrinkage evolution curves of the two under different curing temperatures. The results are relatively consistent among the three measurement channels, with the coefficient of variation (COV) in each data set less than 8.4%. All tests results show that the inflection point appears earlier at higher curing temperature. This is mainly due to the higher hydration rate at higher curing temperature, which in turn causes the cement to form a network structure earlier, and ultimately inhibits external shrinkage. Under the same curing temperature, theinflection time of cement slurry with different w/c is not much different. The results suggest that the effect of curing temperature on the inflection time is significantly higher than w/c.

3.4. Discussions of test results

Comparing the data in Table 4 and Table 3, one can find an interesting phenomenon that the final setting time and the inflection time correlate positively for the same slurry formulation at the same curing temperature. Figure 9 more intuitively shows the relationship between the average inflection time and the final setting time. It can be seen that the setting time of the slurry with w/c=0.3 is approximately equal to the inflection time, while the setting time of the slurry with w/c=0.5 is slightly longer than the inflection time. The correlations between the two parameters are approximately linear for both w/c ratios, suggesting that one may use one data set to estimate the other one. Setting and linear shrinkage of cement are both closely related to the degree of cement hydration. According to the definition of final setting, cement slurry is converted from plastic fluid to solid after a period of hydration. Assuming the cement can continuously imbibe water from the environment to remain fully saturated at all times, theoretically there would be no external bulk shrinkage. However, under sealed condition, the internal volume reduction as a result of chemical shrinkage will exert a vacuum force on the solid skeleton of the cement mixture, leading to autogenous shrinkage. This study revealed that the autogenous shrinkage of the oil well cement slurries investigated here is approximately zero, suggesting the vacuum force is too small to generate any appreciable strain on the set cement.

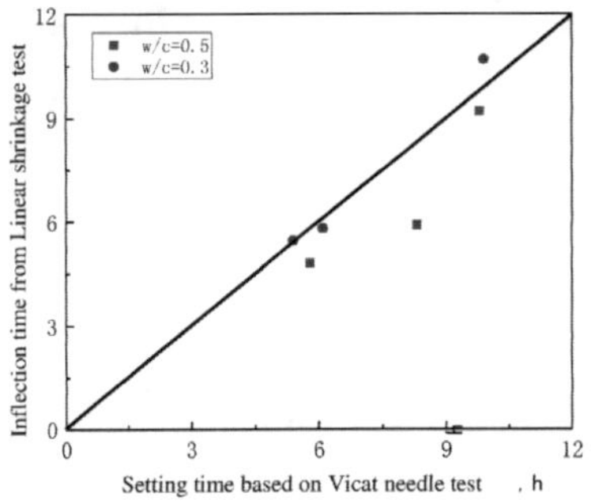

Figure 9. The relationship between inflection point time and final setting time.

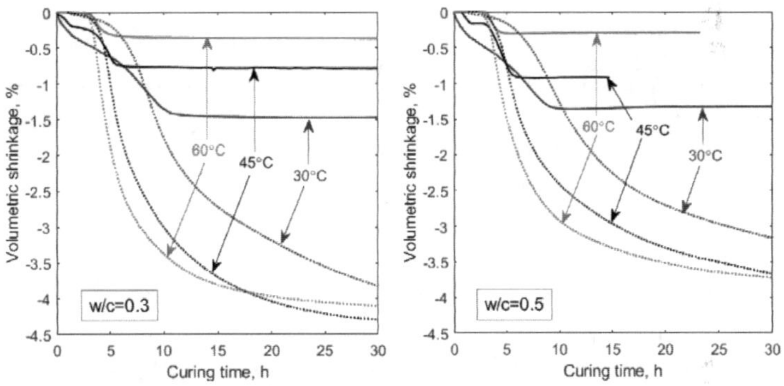

Figure 10. The effect of curing temperature and w/c on cement chemical shrinkage (dotted lines) and bulk shrinkage (solid lines) of oil well cement (bulk shrinkage is calculated based on linear shrinkage test results assuming uniform shrinkage in all directions).

The corrugated tube test measures the cement bulk shrinkage in one dimension, i.e. linear shrinkage. If one assumes the linear shrinkage is the same in all directions during the entire test period, then the volumetric bulk shrinkage can be deduced from the linear shrinkage. Figure 10 compares the calculated volumetric bulk shrinkage with the chemical shrinkage test results of the two cement slurries with different w/c at various curing temperatures. As discussed earlier, chemical shrinkage is the total shrinkage associated with cement hydration reaction, therefore, it is the maximum possible volumetric shrinkage. At lower curing temperatures (30°C and 45°C), it is observed that the calculated bulk shrinkage value exceeds the chemical shrinkage during the early hydration period. The overestimation is likely caused by the fact that linear shrinkage is not equal in all directions and may primarily happen in the measurement (axial) direction during early stage at lower temperatures. This may happen when the corrugated tube is slightly overfilled, resulting in small tensile stresses in the corrugated tube and compressive stresses in the cement slurry; as the cement shrinks, the stresses are gradually released and deformation primarily occurs in the axial direction. At higher temperatures, the corrugated tube is greatly softened, which can also lead to stress release and the observed shrinkage in the axial direction would be much less. In any case, the calculated bulk shrinkage eventually becomes much less than the chemical shrinkage. The crossover generally occurs at or before the final setting time of the cement.

Using the compound composition of the cement presented in Section 2.1 along with the empirical function proposed by Pang et al. [8], the ultimate chemical shrinkage of the cement at fully hydrated condition under 25°C curing temperature may be estimated. The ultimate chemical shrinkage at other temperatures may be estimated by assuming a linear reduction rate of 0.6%/°C with the increase of temperature [6]. Based on the actual chemical shrinkage test results obtained in this study, the degree of hydration of the cement at the initial and final setting time may be derived. Table 5 shows the calculated degree of hydration corresponding with the time of set at different curing temperatures. It appears that the degree of hydration at both initial set and final set increases significantly with increasing curing temperature. A previous study has reported that the degree of hydration of Class H cement with w/c=0.35 is approximately 3.88±0.38% at the initial setting time under different curing temperatures (from 10 to 60°C) [24]. The different values obtained here may be associated with different experimental methods as well as different cement compositions.

Table 5. Calculated degree of hydration at time of set under different curing temperatures.

Temperature/°C	CS_0/(mL/100g)	w/c=0.5		w/c=0.3	
		α_I /%	α_F /%	α_I /%	α_F /%
30	5.01	10.8	19.9	10.6	22.6
45	4.56	25.3	37.9	9.2	24.8
60	4.15	28.8	37.1	21.9	32.5

[a] α_I is the calculated degree of hydration at the time of initial set;
[b] α_F is the calculated degree of hydration at the time of final set;
[c] CS_0 is the estimated ultimate chemical shrinkage at complete hydration.

4. Conclusions

The influences of w/c ratio and curing temperature on the shrinkage behavior of Class G oil well cement were investigated using chemical shrinkage and corrugated tube test methods. The following conclusions may be drawn from this study:

(1) The Vicat setting time of cement slurries with both w/c ratios (0.3 and 0.5) decreases with the increase of curing temperature; Under the same curing temperature, the setting time of the cement slurry with lower w/c is generally slightly shorter than that with higher w/c.

(2) The linear shrinkage of cement measured by the corrugated tube method is relatively high during the plastic stage (i.e. before setting), but reduces to near zero during the autogenous shrinkage stage (i.e. after setting); the plastic shrinkage measured by corrugated tube method decreases significantly with increasing curing temperature.

(3) The significant change in linear shrinkage rate of cement at time of set produced an obvious inflection point on the shrinkage test curve, which correlates very well with the setting time measured by Vicat needle test method especially for lower w/c ratio. The w/c ratio seem to have very little influence on the time of the inflection point.

(4) The degrees of cement hydration at initial and final setting time are estimated to be about 10% and 20%, respectively, at 30°C based on chemical shrinkage test results, which seem to increase significantly with increasing curing temperature.

Acknowledgements

The authors gratefully acknowledge the financial support provided by National Natural Science Foundation of China (Grant No. 51974352).

References

[1] Smith D K 1990 Cementing, SPE Monograph series 4
[2] Reddy B R, Xu Y, Ravi K, et al 2007. *Drilling & completion* **24(1)** 104-114

[3] Parcevaux P A, Sault P H 1984. SPE 13176 Annual technical conference and exhibition 1-7
[4] Justnes H, Van Loo D, Reyniers B, et al 1995. *Adv. Cem. Res.* **7** 85-90
[5] Lyomov S K, Backe KR, Skalle P, et al 1997. *The petroleum society* 1-7
[6] Pang X, Bentz D P, Meyer C, et al 2013. *Cem. Concr. Compo* **39** 23-32
[7] Pang X, Meyer C, Darbe R, et al 2013. *ACI Mater* **110(2)** 137-148
[8] Pang X, Meyer C 2012. *ACI Mate*r **109(3)** 341-352
[9] API TP 1997, Shrinkage and Expansion in Oilwell Cements, first edition, American Petroleum Institute, 10TR2 (58 pp.)
[10] ANSI/API Recommended Practice 2015 , Recommended Practice on Determination of Shrinkage and expansion of Well Cement Fromulations atmospheric Pressure, first edition, American Petroleum Institute, reaffirmed, 10B-5(13 pp.)
[11] Lura P, Jensen O M, Van Breugel K 2003. *Cem. Concr. Res.* **33(2)** 223-232
[12] Jensen O M, Hansen P F 2001. *Cem. Concr. Res.* **31** 1859-1865
[13] Lura P, Jensen O M 2007. *Mater. Struct.* **40(4)** 431-440
[14] ASTM 2019, Standard Test Method for Autogenous Strain of Cement Paste and Mortar, ASTM International, C1698-19 (8 pp.)
[15] Snoeck D, Jensen OM, De Belie N 2015. *Cem. Concr. Res* **74** 59-67
[16] Bao Y, Meng W, Chen Y, et al 2015. *Mater. Lett* **145** 344-346
[17] Tian Q, Jensen O M 2009. *Journal of the Chinese Ceramic Society* **37(1)** 39-45
[18] Brooks J J, Cabrera J G 1998 Factors affecting the autogenous shrinkage of silica fume high strength concrete. In: Proceedings of international workshop on autogenous shrinkage of concrete, Hiroshima, Japan. Japan: Japan Concrete Institute; 185-192
[19] Chu I, Kwon S H, Amin M N, et al 2012. *Constr Build Mater.* **35** 171-182
[20] ASTM C 2019. Standard Test Methods for Time of Setting of Hydraulic Cement by Vicat Needle, ASTM International, 191-19 (8 pp.)
[21] API Recommended Practice 2013, Recommended Practice for Testing Well Cements, second edition, American Petroleum Institute, 10B-2(111 pp.)
[22] Pang X, Jimenez W C, Iverson B J 2013. *Cem. Conc. Res.* **54** 69-76
[23] Wyrzykowski M, Hu Z, Ghourchian S, et al 2017 Mater. *Struct.* **50** 57
[24] Zhang J, Weissinger E A, Peethamparan S, et al 2010 *Cem. Conc. Res.* **40(7)** 1023-1033

Weak inter-layer bonding in extrusion 3D concrete printing: a comparative analysis of mitigation techniques

Nazar Muhammad Salman[1,*], Guowei Ma[1,2], Nauman Ijaz[1] and Li Wang[2]

[1] College of architecture and civil engineering, Beijing university of technology, 100 Pingleyuan, Chaoyang district, Beijing 100124, China.
[2] School of civil and transportation engineering, Hebei university of technology, 5340 Xiping road, Beichen district, Tianjin 300401, China.

*Email: msn.salman@gmail.com

Abstract. Recently, there has been meaningful progress towards the construction three-dimensional printing (3DP) of concrete (3DCP). However, weak inter-layer bonding (ILB) in extrusion 3DCP is a constraint that can critically influence the mechanical capacities, the stability of the structures, and durability. Various techniques have been proposed to mitigate weak ILB. Though, these techniques do not systematically address the basic causes of weak ILB and display inconsistency of the results. Additionally, the scope of the testing has been rather limited. Here, an approach for the mitigation of weak ILB is proposed, as part of ongoing research. It is based on stabilizer mortars (SMs) of varying mix compositions, according to the printing time interval (TI). This study compares mitigation techniques based on the statistical analysis (Chebyshev theorem, T-test, and standard error) of the test data against a 90% confidence interval. The proposed technique demonstrated peak mitigation of weak ILB with less variation in test results at extended TI up to 120 min. Besides efficacy, the simplicity of the technique i.e. practicality of the method and the materials cost, availability, safety, and sustainability remain ideal.

1. Introduction

Three-dimensional concrete printing (3DCP) has been gaining attention for large-scale construction considering technical, economical, and environmental aspects in recent years [1-3]. The 3DCP techniques are principally based on powder bed deposition and extrusion. The extrusion 3DCP techniques seem to be more practical and popular based on economic feasibility and overall level of technological maturity [4, 5]. Extrusion 3DCP techniques may be centered on ordinary Portland cement (OPC) or geopolymer concrete [6, 7]. The defining feature of extrusion 3DCP is the automatic deposition of layers to assemble structures. The layers of 3DP concrete are connected at their surfaces by adhesion [8, 9]. Due to printing time intervals (TI) between depositions i.e. time gap effects, the resulting layer interfaces may develop severe weakness [10, 11]. The inter-layer bond (ILB) strength decreases due to a lack of intermixing because of changes in material properties, and due to effects induced by printing process parameters [12-14]. Weak ILB in 3DP concrete is a constraint that can critically influence the mechanical capacities, the stability of the structures, and durability [3, 11, 12]. Previous studies are focused on weak ILB in OPC based extrusion 3DCP, whereas weak ILB is also an issue in geopolymer based extrusion 3DCP [15].

The ILB strength is determined through the measurement of the inter-layer tensile bond strength (ILTBS) and the inter-layer-shear bond strength (ILSBS). The ILSBS is of greater practical importance in extrusion 3DCP because it is directly related to the stability of the structures [16]. Currently, there is no standard testing approach for measuring both the ILTBS and the ILSBS in 3DP concrete [9]. Le et al first investigated the weak ILB in 3DP concrete [17]. The study showed reduced ILTBS, compared to material tensile strength (i.e. 2.3 MPa to 0.7 MPa with respect to TI of 15 min and 7 days) with coefficients of variation of 5 to 30%. Moreover, beyond the TI of 90 min, the reduction in ILB is insignificant. In another study, the ILTBS was observed to decrease by 9.9%, 14.1%, and 23.1%, at TI of 2 min, 10 min, and 1 day, respectively [18]. It is pertinent to mention here that there is a scarcity of literature focusing on ILSBS which is an extremely critical aspect of structural stability.

Researchers have proposed various techniques for the mitigation of weak ILB. However, these techniques do not comprehensively address the basic causes of weak ILB. There is a high variation in the test results (up to 45%) and a lack of consistency. Additionally, the scope of the testing was rather limited. Either these studies failed to simultaneously determine both the ILTBS and the ILSBS or the testing was limited to a TI of 15 min [9, 12, 19]. It is hypothesized that effective mitigation of weak ILB at TI exceeding 30 min would be compromised. Therefore, the authors envisaged studies focusing on mitigation of both ILTBS and ILSBS considering extended TI.

This study aims to develop a weak ILB mitigation technique for extrusion 3DCP and compare it with other techniques documented in the literature based on detailed statistical analysis. The statistical analysis is centered on the Chebyshev theorem, T-test, and standard deviation. This is to validate the proposed technique that is comprehensive in the scope of testing, effective, and displays the consistency of the results. This study will facilitate the acceptance of extrusion 3DCP techniques as a regular construction practice.

2. Background: Techniques for the mitigation of weak interlayer bonding (ILB) in OPC-based 3DCP

This section discusses the background of different weak ILB mitigation techniques documented in the literature considered for comparison purpose in this study:

Contour crafting (CC) is a form of extrusion 3DCP. An attempt was made to mitigate weak ILB through the interlocking of the layers. The ILB was reported to have been enhanced by an average of 26% [9]. CC utilizes a trowel control mechanism to shape the layers into complex geometries. Therefore this method may not be practicable across the full spectrum of extrusion 3DCP.

Ehsan et al proposed the use of a sulfur-black carbon-sand (SBCS) composite polymer mortar for enhancing ILB [19]. This study did not determine the mitigation of weak ILB with respect to TI and the results showing enhancement in ILSBS were based on simulated molecular dynamics between the CSH and the SBCS composite. Since this method relies on heating for the manufacture of the SBCS mortar so it may not be feasible for on-site large-scale construction.

Taylor et al proposed mitigation of weak ILB by effective bond area amplification through the application of OPC-based pastes at the interfaces [12]. Four pastes of different compositions were tested i.e. paste-control, paste-retarder, paste-viscosity modifying agent, and paste-superplasticizer. In this study, the TI was fixed at 15 min and the ILSBS was not determined.

3. Materials and methods

The proposed technique is based on the application of stabilizer mortars (SMs) at the interfaces through extrusion on to the surfaces of substrate layers during printing.

3.1. Selection of materials for the SMs

The binder component of the SMs is a blend of OPC and high belite calcium sulfoaluminate (CSA) cement. Cellulose fiber (CF) pulps are incorporated as internal curing (IC) agents. The CF retains water for IC, aids hydration and mechanical anchorage, and helps to reduce shrinkage. The CSA cement consumes excess surface water, acts as a binder for the CF, and permits matching of the elastic modulus between the overlaying and substrate concrete layers.

3.2. *Mix proportions of the SMs*

The SMs mixes were designed according to TI i.e. 60 min, 90 min, and 120 min. Therefore, the proportion of the CSA cement in the blends was increased consecutively; however, it was restricted by the setting time of the blends [20, 21]. Three (03) conservative water/ binder (w/b) ratios i.e. 0.45, 0.47, and 0.49 were selected to ensure the utilization of excess water from the surfaces of the fresh 3DP concrete layers [22]. The SM mixes were designated SM-1 through SM-12. Control specimens being exactly similar to test specimens but without SMs, those were designated SMC, SMC-60, and SMC-90 respectively. For SMC specimens the TI is taken at $\sim 3\ min$.

3.3. *Methods*

3.3.1. Consistency of the SM. The effectiveness of the SMs is based on their flowability [12, 21, 23]. The consistency was quantified by the mini-slump test [24].

3.3.2. Printing of test specimens. SM coatings ~1 mm were extruded onto the top surfaces of 3DP concrete layers through a manual extrusion device immediately after extrusion. Then overlaying layers were printed at TI of 60 min, 90 min, and 120 min. The specimens were cured under ambient conditions for 28 days prior to testing.

4. Results and discussion

4.1. *Comparative analysis of weak ILB mitigation*

4.1.1. Comparative analysis of weak ILTBS mitigation A comparative statistical analysis was performed on ILTBS test data from four studies including one pertaining to geopolymer 3DP concrete, as presented in Figure 1 (a). The comparative statistical analysis of the weak ILTBS mitigation test data from two studies (TI of 15 min) mentioned (in Section 2), is presented in Figure 1 (b).

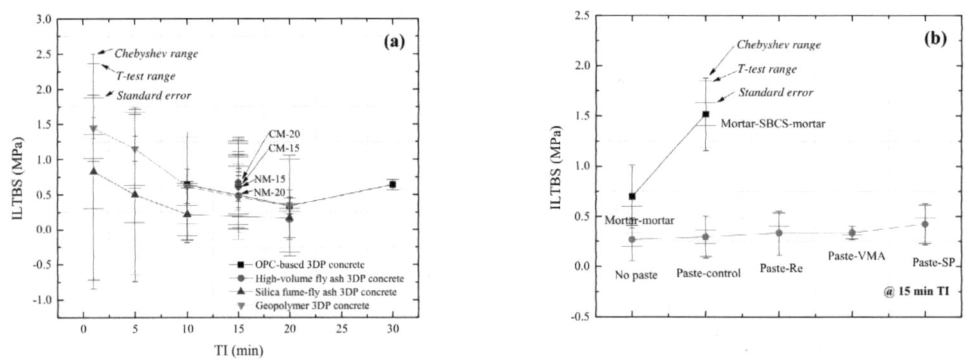

Figure 1. ILTBS; (a) statistical analysis of comparative ILTBS test data; (b) comparative statistical analysis of weak ILTBS mitigation test data, based on [11], [14], [15], [12], and [19].

From Figure 1 (a), it can be seen that overall, ILTBS varies from ~1.4 MPa to about 0.25 MPa, between TI of 1 min to 30 min, with a coefficient of variation of ~40%. The two mitigation techniques enhanced ILTBS by 59% and 100% i.e. to 0.43 MPa and 1.5MPa respectively for a TI of 15 min [12, 19]. The ILTBS mitigation results show a variation of ~20%. Statistical analysis of the mitigation of weak ILTBS at 60, 90, and 120 min TI, based on test data of the proposed technique is presented in Figure 2. At each TI the optimal mixes i.e. SM-2, SM-6, and SM-10 have been considered.

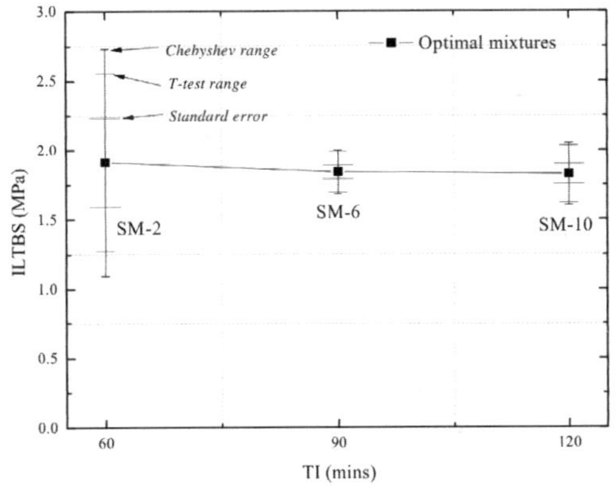

Figure 2. Statistical analysis of weak ILTBS mitigation test data through the use of SMs.

At 60 min TI, the ILTBS was enhanced to 1.91 MPa i.e. marginally exceeding that of the control (SMC), while exceeding SMC-60 (i.e. control at 60 min TI) by approximately 32%. At 90 min TI, the ILTBS was measured at 97% of SMC. At 120 min TI, the ILTBS was enhanced to 96% of SMC and by 185%, as compared to SMC-120. Based on standard error, the test results show a 20-45% variation. Overall, this technique offers superior weak ILTBS mitigation, as compared to the other discussed techniques.

4.1.2. Comparative analysis of weak ILSBS mitigation Statistical analysis was performed on ILSBS test data from one study, as presented in Figure 3 (a). The comparative statistical analysis of weak ILSBS mitigation simulation data from the study discussed (in Section 2), is presented in Figure 3 (b).

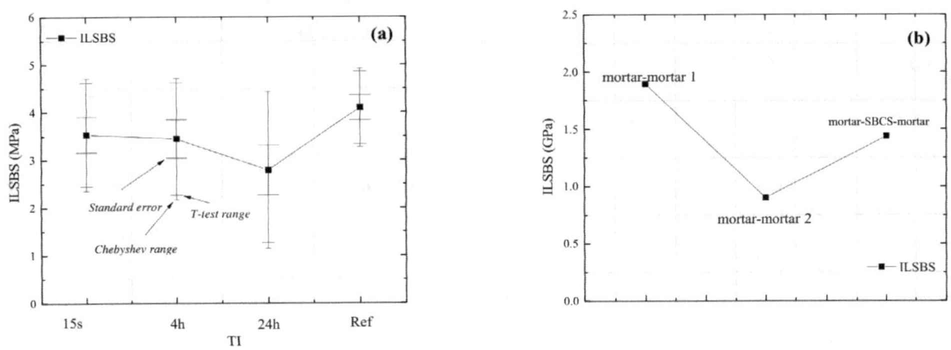

Figure 3. ILSBS; (a) statistical analysis of ILSBS test data; (b) statistical analysis of weak ILSBS mitigation simulation data, based on [16], and [19].

Figure 3 (a) shows that ILSBS varies from ~3.6 MPa to 2.60 MPa between TI of 15 sec to 24 h, with a coefficient of variation of about 25%. Based on simulation results (Figure 3 (b)), the ILSBS has been enhanced to 1.4 GPa [19]. Therefore, the validity of the simulation data must be confirmed by experimentation. Furthermore, the mitigation of weak ILSBS was not considered with respect to TI. Statistical analysis of the mitigation of weak ILSBS at 60, 90, and 120 min TI, based on test data of the proposed technique is presented in Figure 4.

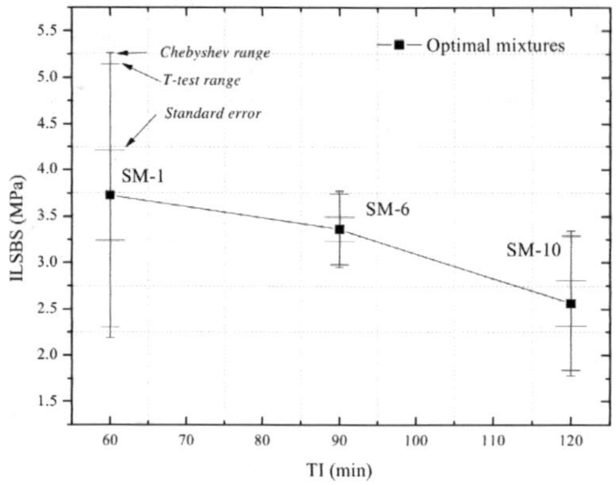

Figure 4. Statistical analysis of weak ILSBS mitigation test data through the use of SMs.

At 60 min TI, the ILSBS was measured to marginally exceed that of SMC by 5%, and exceeding SMC-60, by approximately 40%. At 90 min TI, the ILSBS was measured at 94% of SMC. At 120 min TI, the ILSBS was enhanced to 72% of SMC and by 181%, as compared to SMC-120. Based on standard error, the test results show ~30% variation. Again this technique offers effective weak ILSBS mitigation.

5. Conclusion and recommendations

The proposed technique has significantly improved the ILTBS and ILSBS in comparison to other mitigation techniques discussed in this study with special attention to extended TI. At a TI of 120 min, the proposed technique enhanced ILTBS to within 4% of the control strength. The ILSBS was enhanced by 181%, as compared to the control at TI of 120 min. The statistical analysis of the other, weak ILB mitigation techniques show high variation in the test data against a 90% confidence interval; this could seriously question their practical implementation. The proposed technique exhibits greater homogeneity of test data and the validity of the test procedure**.** Simple manufacturing, the use of cheap, safe, and sustainable materials, plus possible additional benefits, marks this technique as particularly feasible for OPC based extrusion 3DCP.

Based on the evaluated data, the following recommendations are drawn:

There is an urgent need to develop standard testing approaches for the measurement of ILB i.e. both ILTBS and ILSBS in 3DP concrete.

Techniques for the mitigation of weak ILB in geopolymer based extrusion 3DCP must be proposed.

References

[1] Lu B, Weng Y, Li M, Qian Y, Leong K F, Tan M J, Qian S 2019 A systematical review of 3D printable cementitious materials *Construction and Building Materials* **207** 477-490.

[2] Wangler T, Roussel N, Bos F P, Salet T A M, Flatt R J 2019 Digital Concrete: A Review *Cement and Concrete Research* **123** 105780.

[3] Schutter G De, Lesage K, Mechtcherine V, Nerella V N, Habert G, Agusti-Juan I 2018 Vision of 3D printing with concrete — Technical, economic and environmental potentials *Cement and Concrete Research* **112** 25-36.

[4] Mechtcherine V, Nerella V N, Will F, Näther M, Otto J, Krause M 2019 Large-scale digital concrete construction – CONPrint3D concept for on-site, monolithic 3D-printing, *Automation in Construction* **107** 102933.

[5] Buswell R A, Leal de Silva W R, Jones S Z, Dirrenberger J 2018 3D printing using concrete extrusion: A roadmap for research *Cement and Concrete Research* **112** 37-49.

[6] Voney V, Landrou G, Odaglia P, Jipa A, Agustí-Juan I, Dillenburger B, Habert G 2018 Geopolymers as inorganic binder in 3D printing application in construction, *Alkali-Activated Materials and Geopolymers: Versatile Materials Offering High Performance and Low Emissions*, Tomar, Portugal, May 27.

[7] Zhang D W, Wang D m, Lin X Q, Zhang T 2018 The study of the structure rebuilding and yield stress of 3D printing geopolymer pastes *Construction and Building Materials* **184** 575-580.

[8] Zhang J, Wang J, Dong S, Yu X, Han B 2019 A review of the current progress and application of 3D printed concrete *Composites Part A: Applied Science and Manufacturing* **125** 105533.

[9] Zareiyan B, Khoshnevis B 2017 Effects of interlocking on interlayer adhesion and strength of structures in 3D printing of concrete *Automation in Construction* **83** 212-221.

[10] Marchment T, Sanjayan J G, Nematollahi B, Xia M 2019 Chapter 12 - Interlayer Strength of 3D Printed Concrete: Influencing Factors and Method of Enhancing, in: J.G. Sanjayan, A. Nazari, B. Nematollahi (Eds.), 3D Concrete Printing Technology, *Butterworth-Heinemann* 241-264.

[11] Sanjayan J G, Nematollahi B, Xia M, Marchment T 2018 Effect of surface moisture on inter-layer strength of 3D printed concrete *Construction and Building Materials* **172** 468-475.

[12] Marchment T, Sanjayan J, Xia M 2019 Method of enhancing interlayer bond strength in construction scale 3D printing with mortar by effective bond area amplification *Materials & Design* **169** 107684.

[13] Tay Y W D, Ting G H A, Qian Y, Panda B, He L, Tan M J 2019 Time gap effect on bond strength of 3D-printed concrete *Virtual and Physical Prototyping* **14(1)** 104-113.

[14] Panda B, Noor Mohamed N A, Paul S, Bhagath Singh G V P, Tan M J, Šavija B 2019 The Effect of Material Fresh Properties and Process Parameters on Buildability and Interlayer Adhesion of 3D Printed Concrete *Materials*.

[15] Panda B, Paul S, Noor Mohamed N A, Tay Y W D, Tan M J 2017 Measurement of tensile bond strength of 3D printed geopolymer mortar *Measurement* **113** 108-116.

[16] Wolfs R J M, Bos F P, Salet T A M 2019 Hardened properties of 3D printed concrete: The influence of process parameters on interlayer adhesion *Cement and Concrete Research* **119** 132-140.

[17] Le T T, Austin S A, Lim S, Buswell R A, Law R, Gibb A G F, Thorpe T 2012 Hardened properties of high-performance printing concrete *Cement and Concrete Research* **42(3)** 558-566.

[18] Nerella V N, Hempel S, Mechtcherine V 2019 Effects of layer-interface properties on mechanical performance of concrete elements produced by extrusion-based 3D-printing *Construction and Building Materials* **205** 586-601.

[19] Hosseini E, Zakertabrizi M, Korayem A H, Xu G 2019 A novel method to enhance the interlayer bonding of 3D printing concrete: An experimental and computational investigation *Cement and Concrete Composites* **99** 112-119.

[20] Marchon D, Kawashima S, Bessaies-Bey H, Mantellato S, Ng S 2018 Hydration and rheology control of concrete for digital fabrication: Potential admixtures and cement chemistry *Cement and Concrete Research* **112** 96-110.

[21] Li W, Yu J, Ma S, hu Y, Ge D, Shen X 2018 The properties and hydration of Portland cement containing calcium sulfoaluminate cement *Ceramics–Silikáty* **62(4)** 364-373.

[22] Trauchessec R, Mechling J M, Lecomte A, Roux A, Le Rolland B 2015 Hydration of ordinary Portland cement and calcium sulfoaluminate cement blends *Cement and Concrete Composites* **56** 106-114.

[23] Marchment T, Sanjayan J G 2019 Method of Enhancing Interlayer Bond Strength in 3D Concrete Printing In: RILEM International Conference on Concrete and Digital Fabrication Springer, *Cham* 148-156.

[24] Gao J, Fourie A 2015 Spread is better: An investigation of the mini-slump test *Minerals Engineering* **71** 120-132.

Effect of various testing parameters on the experimental evaluation of oil well cement cured under simulated downhole conditions

Jiankun Qin[1], Xueyu Pang[1,2,*], Zhongfei Liu[3], Zhengqing Ai[3] and Ning Li[3]

[1] School of petroleum engineering, China university of petroleum (east China), Qingdao 266580, China;
[2] Key laboratory of unconventional oil & gas development (China university of petroleum (east China)), ministry of education, Qingdao 266580, China;
[3] Oil and gas engineering research institute, petroChina tarim oilfield company, Korla 841000, Xinjiang, China.

[*]Email: x.pang@upc.edu.cn

Abstract. It is well known that regular oil well cement systems experiences strength retrogression at curing temperatures above 110ºC and silica sand is typically added to mitigate strength reduction. However, due to the complex experimental procedures associated with high temperature and high pressure curing of oil well cement specimens in the laboratory, historical test data tend to suffer significant uncertainties. A comprehensive study of the various influencing factors on the physical and mechanical property test results of oil well cement system cured under 200ºC is conducted here. The specific influencing factors investigated include curing time (from 14d to 69d), depressurization rate, curing pressure (50MPa vs 150MPa), specimen size (diameter of 25mm vs 50mm) and curing autoclave design. The property of the set cement evaluated include strength and modulus, permeability, indentation modulus and hardness, etc. Test results suggest bigger specimen size and faster depressurization rate may lead to significant specimen damage, causing experimental artifact. Different autoclave design may also lead to significant differences in the mechanical property of the set cement even under the same curing temperature and pressure.

1. Introduction

Oil well cement is a special type of Portland cement. The most commonly used oil well cement are high-sulfate-resistant grade Class G and Class H cement, whose main differences from ordinary Portland cement (OPC) are their low content of tricalcium aluminate (C_3A) [1]. Similar to OPC, the hydration products of oil well cements at low temperatures (<100°C) are mainly C-S-H gel and Ca $(OH)_2$. These hydration products have a compact structure and good mechanical properties. At temperatures above 110°C, C-S-H gel is gradually converted to crystalline phases such as α-C_2SH, causing the mechanical properties of the set cement to degrade and this phenomenon is known as strength retrogression. Silica flour is mixed with oil well cement to decrease the calcium oxide to silicon dioxide (Ca/Si) ratio of the mixture, which is desirable for the formation of stable non-retrogressing phases at high temperatures, such as 11 Å tobermorite and xonotlite [2-4].

Due to the need to simulate the high pressure and high temperature (HPHT) condition in the downhole environment of oil wells, the preparation of set oil well cement specimens for property evaluation is a highly complex process [1, 5]. It typically consists of the following main steps: 1) casting the cement slurry in curing molds; 2) put the slurry-filled molds in a HPHT autoclave; 3) ramp the temperature and pressure of the autoclave to target values in a pre-specified period of time; 4) hold the temperature and pressure constant for the duration of the test, 5) ramp the temperature and pressure of the autoclave to the ambient condition; 6) take out the specimens and remove the molds. Sometimes, in order to simplify the experimental procedures or to simulate field condition of steam injection wells, the cement may be allowed to set outside of the autoclave at lower temperatures and demolded before being subjected to HPHT curing. These complicated specimen preparation procedures can bring significant uncertainties in the experimental results, which may subsequently lead to experimental artifact. For example, it is very well known that the depressurization rate employed at the end of the HPHT curing has a very important impact on the integrity of the specimens [6]. To further complicate the situation, the severity of such effect may be dependent on cement mixture composition and specimen size [7]. In extreme cases, the specimens could be completely disintegrated up on taking out of the molds [6]. Iverson [8, 9] et al. performed a comprehensive study regarding the influences of silica dosage (40% to 80%), curing temperatures (177°C to 343°C) and curing time (14 d to 180 d) on the physical and mechanical properties of oil well cement formulations. However, test data were found to be highly scattered and no obvious trend of strength change with silica dosage, curing temperature, or curing time could been found. Therefore, it is important to study the influences of various testing parameters on the experimental evaluation results of oil well cement.

Although oil well cement specimens are generally cured under simulated downhole environment, the evaluation of their physical and mechanical properties are typically performed at the ambient condition. In order to eliminate depressurization damage to the cement specimens as well as to obtain their "in-situ" properties under HPHT environment, significant efforts have been made to develop customized testing equipment that may cure and test oil well cement mixtures within the same device [10, 11]. However, until now, the only commercialized in-situ test device of such type is based on measurement of ultrasonic transit time (i.e. the Ultrasonic Cement Analyzer, or UCA) [12, 13]. The advantage of UCA test is its ability to evaluate the physical property of cement continuously under HPHT environment, while its disadvantage is the measured acoustic property can only be used to calculate compressive based on empirical equations, which can be highly inaccurate. Reddy [12] et al. conducted UCA tests on oil well cement cured under HPHT conditions and found that the heating rate during curing had a great influence on the sonic strength development of set cement. As the employment of different cement curing chambers may lead to differences in heating rate, their influences on the properties of set cement should also be studied.

Unlike curing temperature, curing pressure is generally believed to have only a minor impact on the mechanical property development of oil well cement systems. It is believed that curing pressure will not change the chemical composition of cement hydration products and hence its effect on the mechanical properties is primarily from compaction. Early research [14] has showed that increasing curing pressure up to 13.8MPa would improve the shorter term compressive strength of cement, while further increases in curing pressure has little effects. More recently, it has been found that curing pressure has qualitatively similar, but quantitatively much smaller effects than curing temperature in terms of accelerating the hydration reaction rate of oil well cement [15-19]. The influences of curing pressure in a much higher pressure range (50MPa-150MPa) will be studied here to improve our understanding in this area.

In summary, due to the lack of sufficient experimental data as well as the complex influencing factors on experimental results, the long-term performance of silica-enriched oil well cement systems is not fully understood. The goal of this study is to investigate the influences of various testing parameters on the physical and mechanical performance of oil well cement systems. The testing parameters studied include depressurization rate, temperature rising rate, specimen size, curing pressure, and curing time. Several previously optimized cement formulation with 70%-90% by weight of cement (bwoc) total silica addition were cured for different durations varying from 14 d to 69 d. The specimens were directly cured in HPHT chambers with molds to simulate the cementing of ultradeep wells and geothermal wells. The properties evaluated include compressive strength, elastic modulus, Poisson's ratio and permeability.

2. Experimental procedure

2.1. Raw materials

Aksu Class G oil well cement (3.127 g/cm^3, D50 19.83 μm), silica fume (2.332 g/cm^3, D50 0.37 μm), fine silica flour (2.626 g/cm^3, D50 15.52 μm) and coarse silica flour (2.666 g/cm^3, D50 53.06 μm) were provided by Tarim Oilfield Company. The specific gravities of the powder materials were tested by a pycnometer (model UltraPYC 1200e) from Quantachrome Instrument, USA. The particle size of powder materials was analyzed by a laser particle analyzer (Malvern Mastersizer 2000). The D50 (i.e. median particle size) of cement and silica flour are of the same order of magnitude, while that of silica fume is 1-2 orders of magnitude lower. Silica fume is known to accelerate the process of cement hydration and enhance the compactness of set cement [20]. X-ray fluorescence (XRF) spectrum analysis results of dry cement and silica fume was shown in

Table 1. Retarder (BCR-300L), fluid loss reducing agent (BXF-200L), dispersant (BCD-210L), defoaming agent (G603), high temperature suspension agent (BDJ-300S) and other oil well cement admixtures were obtained from Tianjin PetroChina Boxing Technology Co., Ltd.

Table 1. X-ray fluorescence spectrum (XRF) test results of silica fume and aksu cement.

Oxide	Al_2O_3	CaO	Fe_2O_3	K_2O	MgO	Na_2O	SO_3	SiO_2	Free lime
Silica fume	0.297	0.597	0.162	0.776	0.322	0.268	0.578	96.697	N/A
Cement	2.985	65.134	6.675	0.805	1.962	0.179	3.15	18.452	1.65

2.2. Mixture proportion and preparation of samples

Four different cement formulations were designed as shown in Table 2. The dosages of different components are expressed in % by weight of cement (bwoc). The preparation of cement sample consists of four steps: mold preparation, slurry mixing, HPHT curing in water-saturated condition, specimen cutting and grinding. All final test samples were 25×50mm cylinders except for the test series that studies the influences of specimen size on experimental results.

Table 2. Cement formulations Z1 to Z4 in % bwoc (mix volume is 500mL per slurry).

Formula	Cement	fine silica	coarse silica	Silica fume	BCJ-300L	BCD-210L	BCR-300L	Water
Z1	100	15	50	5	4	5.5	5	49.6
Z2	100	30	40	0	4	5.5	4.5	50.6
Z3	100	40	40	0	3.75	6	4	54.14
Z4	100	50	40	0	3.5	6.5	4	57.01

*The dosages of BXF-200L and G603 in each formulation are 6% bwoc and 0.5% bwoc, respectively.

2.3. Test methods

(1) Mechanical property test

UTM5105X microcomputer-controlled electronic universal testing machine from the Shenzhen Sansi Zongheng Company was used for compressive strength test. The testing machine was equipped with a video extensometer (MercuryRT series) from Sobriety S.R.O., Czech Republic, which can simultaneously measure the horizontal and vertical deformation of the sample for calculating Young's modulus and Poisson's ratio.

(2) Permeability test

For water permeability test, samples taken from the curing autoclave were directly used. The applied confining pressure was 5MPa while the applied pressure differential along the sample was about 1.5-2 MPa. The water permeability was calculated according to Formula 1. For gas permeability test, the specimens were dried by vacuum in a freezer dryer at room temperature for one week before testing. Gas permeability of each specimen was measured at four different upstream pressures (0.5-2 MPa) with a constant confining pressure of 4 MPa and calculated using Formula 2. Klinkenberg gas permeability was obtained as the intercept of the plot of gas permeability as a function of the inverse of average applied pressure, i.e. $2/(P_1+P_0)$.

$$K_w = \frac{Q\mu_w L}{A\Delta P} \times 1000 \quad (1)$$

$$K_g = \frac{2Q_0 P_0 \mu_g L}{A(P_1^2 - P_0^2)} \times 1000 \quad (2)$$

where K_g is the gas permeability, mD; P_1 is the inlet pressure for gas permeability measurement, atm; P_0 is atmospheric pressure, which is 1 atm; μ_g is the nitrogen gas viscosity, 0.0178 mPa·s; Q_0 is the volume flow of gas at atmospheric pressure, cm^3/s; L is the sample length, cm; A is the sample cross-sectional area,cm^2; K_w is the water permeability, mD; Q is the water flow rate, cm^3/s, ΔP is the pressure difference for water permeability test, atm; μ_w is the water viscosity, 1 mPa·s.

(3) Nanoindentation test

A Bruker TI Premier nano-indentation instrument with a pressure range of 1-10000μN and an indentation depth of 0-3μm was used for mico-mechanics evaluation of selected specimens. A small piece of sample was impregnated with epoxy resin and subsequently grinded and polished using an automatic grinder-polisher. All tests were programmed in load control mode such that the indenter came into contact with the sample surface with a maximum load of 3500 μN. A total of 400 indents (a matrix of 20 × 20) were conducted per specimen. The spacing between the indents was 10 μm in both directions, resulting in a total indentation area of 200 × 200 μm.

3. Results and discussion

3.1. *Influence of curing time on experimental results*

In the first test series, two HPHT autoclaves from Chandler Engineering (Model 1910) were used to cure the four slurries listed in Table 2 for durations of 14 d and 69 d, respectively. The temperature and pressure profiles during the initial ramp-up are shown in Figure 1 (a). It took approximately 4 hours for the autoclaves to reach the target temperature of 200°C and pressure of 150MPa. The temperature profiles of the two tests were almost overlapped, but the pressure profiles deviated between 150min to 250min due to pressure fluctuations of the14-day-cure test. At the end of the curing period, the autoclaves were switched off to allow depressurization during the process of natural cooling, which took about 1-2 days.

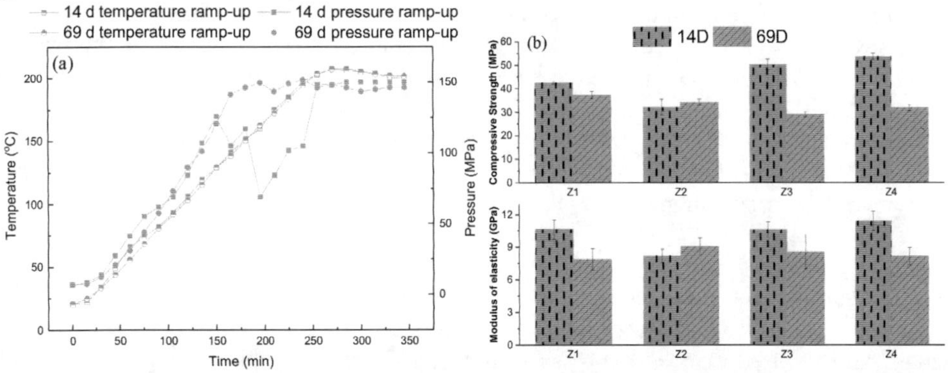

Figure 1. Influence of curing time on experimental results (200°C and 150MPa) (a) Temperature and pressure profile during ramping up (b) Mechanical properties.

The compressive strength and elastic modulus test results of the four formulations cured for 14-day and 69-day are shown in Figure 1(b). It can be seen that with the increase of curing time, the compressive strengths of slurry Z1 and Z2 were little changed, while those of slurry Z3 and Z4 declined significantly (about 41%). The Young's modulus of slurry Z1, Z3 and Z4 apparently decreased over time (about 19-28%) while that of Slurry Z2 slightly increased (~10%). The results also indicate that increasing the dosage of fine silica from 30% to 50% significantly increased the compressive strength and Young's modulus of the set cement during the relatively early age of 14 d, but this also resulted in more severe strength retrogression in the long-term. Slurry Z2 seems to be the optimal formulation for long-term strength stability.

Table 3. Water / Gas permeability of Z1 to Z4.

	Z1	Z2	Z3	Z4
14 d Water permeability (mD)	0.00091	0.00130	0.00091	0.00090
14 d Gas permeability (mD)	0.01900	0.01200	0.02000	0.02500
69 d Water permeability (mD)	0.00260	0.00150	0.00170	0.00170
69 d Gas permeability (mD)	0.03300	0.01900	0.01800	0.01400

Permeability test results of the four formulations are shown in Table 3. Each test result is the average of two specimens. Overall, the repeatability/consistency of both gas and water permeability test results are very good. At 14 d, the water permeability test results showed very little dependence on slurry composition (approximately 0.001mD for all formulations). At 69 d, all formulations showed increased water permeability, with slurry Z1 being the most dramatic and slurry Z2 being the most stable. The gas permeability of all formulations is an order of magnitude higher than the water permeability. This is due to the coarsening effect of the cement microstructure during drying [21].

3.2. Influence of depressurization rate on experimental results

In the second test series, the curing autoclave used was the same as those in the first test series (i.e. Chandler Engineering, Model 1910). The temperature and pressure profiles during the ramp-up period were also similar to the previous tests with a curing period of 14 d. In order to study the effect of the depressurization speed on the experimental test results of the set cement, the temperature and pressure were slowly reduced over a period of 200 hours (or 8 days) by manual control, as shown in Figure 2(a). Figure 2(b) compares the test results of slurries Z1 and Z2 cured for the same duration of 14 days at 200°C and 150MPa, but with different depressurization schemes. Quick depressurization is achieved by natural cooling as described earlier. Comparing the test results, it can be seen that the specimens' water permeability values are dramatically reduced with slow depressurization (65% reduction for slurry Z1 and 46% reduction for slurry Z2), possibly due to reduced specimen damage. At the same time, the compressive strengths are increased with slow depressurization. The amount of increase in compressive strength is 11% for slurry Z1 and 29% for slurry Z2. The increase in Young's modulus and Poisson's ratio of slurry Z2 appear to be consistent with compressive strength. However, no statistically meaningful change is observed for the Young's modulus and Poisson's ratio of slurry Z1. As discussed earlier, quick depressurization may lead to destruction of the sample, which will affect the test results. Test results obtained here seem to suggest that the severity of property deterioration associated with depressurization may also depend on the cement composition.

3.3. Influence of curing pressure on experimental results

In the third test series, the test procedure was similar to the second test series with the exception that the maximum pressure was reduced to 50MPa to study the influence of curing pressure on experimental results. The manual depressurization took about 80 hours, while the temperature reduction scheme were kept the same as the second test series (Figure 3(a)). It can be seen from Figure 3(b) that, with the increase of the curing pressure, the mechanical property of slurry Z1 is apparently increased possibly due to the compaction effect with higher pressure while the mechanical property of slurry Z2 is little changed. Similarly, the water permeability of slurry Z1 is significantly reduced with increasing curing pressure, while that of slurry Z2 is almost constant. These results indicate that the curing pressure has a greater influence on the formulation with silica fume, but had little effect on the formulation without silica fume.

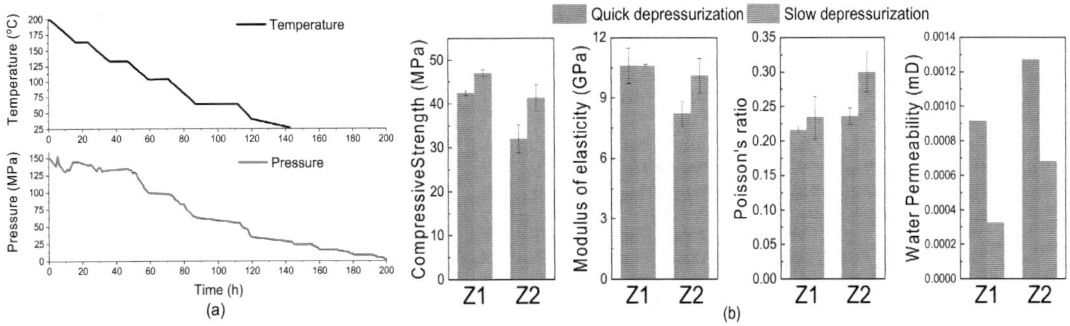

Figure 2. Influence of depressurization rate on experimental results (200°C/150MPa, 14d cure) (a) Temperature and pressure profiles during ramping down (b) Physical and mechanical properties.

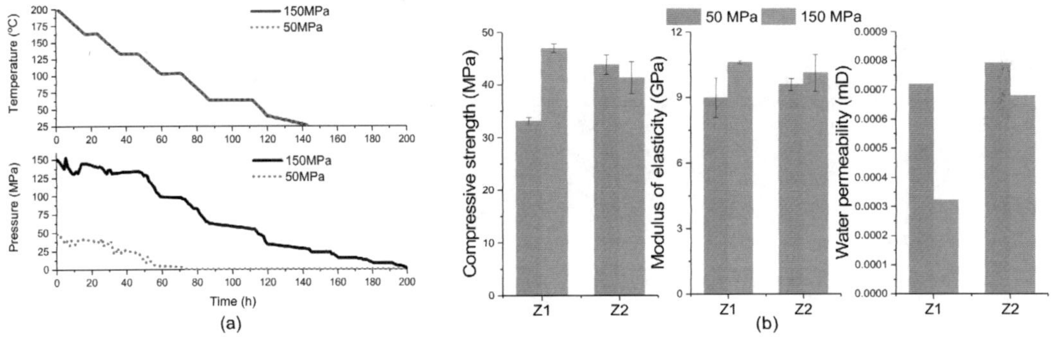

Figure 3. Influence of curing pressure on experimental results (200°C, 14d cure) (a) Temperature and pressure profiles during ramping down (b) Physical and mechanical properties.

3.4. Influence of specimen size on experimental results

Cement specimen damage during depressurization is primarily caused by fast release of pore pressure. The transport path of pore water is strongly dependent on specimen size. Theoretically, for the same depressurization rate, larger specimens would be more susceptible to damage due to longer pore water transport path, which lead to longer equilibrium time and hence larger pressure differential. In the previous two test series, specimens of two different sizes were produced (diameter × height): 25×50mm and 50×100mm. Test results of the smaller specimens were reported previously in section 3.3. The influences of specimen size on compressive strength are shown in Figure 4. Under the curing pressure of 50 MPa, the compressive strength of the big samples was much less than that of the small samples (reduced 48%-64%). However, under the curing pressure of 150MPa, the compressive strength of the two size specimens are relatively close, with a difference of about 15%.

In order to further investigate the causes of strength differences between big and small samples, nano indentation method was used to analyze the micromechanic properties. The test data are presented in Figure 5. For the big sample, the average elastic modulus is 17.35 GPa and the average hardness is 0.70 GPa; for the small sample, the average elastic modulus is 13.19 GPa and the average hardness is 0.45 GPa. Since the elastic modulus and hardness of the big sample on the microscopic scale is higher than those of the small sample (which may be attributed to inhomogeneity of the material), its mechanical property degradation on the macroscopic scale is unlikely to be caused by changes of hydration products. It is suspected that the cracking during depressurization caused the macroscopic property degradation of the big samples. Therefore, using small specimens can potentially help reduce the risk of cracking during the pressure release period after HPHT curing.

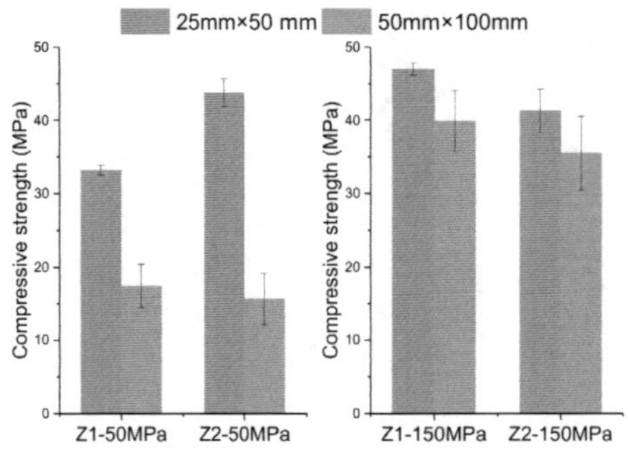

Figure 4. Comparison of 14-day test results of different size samples.

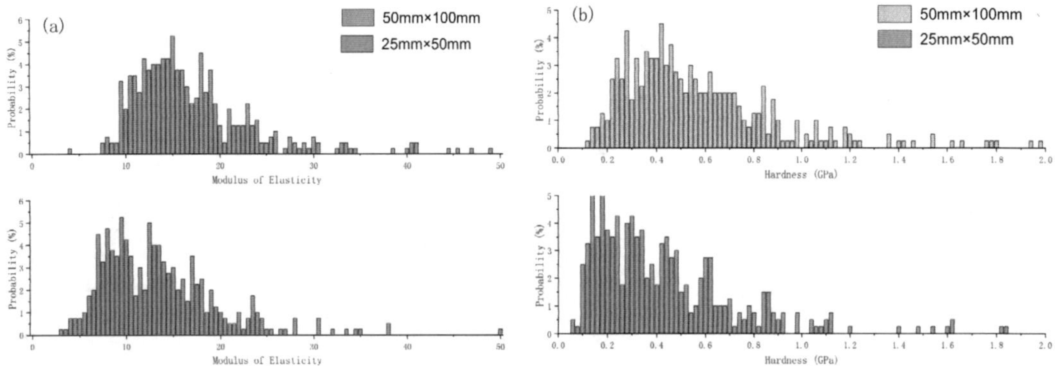

Figure 5. Nanoindentation test results of slurry Z2 (a) elastic of modulus (b) hardness.

3.5. Influence of curing autoclave on experimental results

In the final test series, an additional HPHT autoclave from Jiangsu Hongbo Manufacturing Company was employed to study the influence of different curing chambers on experimental results. The main difference between the curing chambers is the location of the heating elements: they are located inside the curing chamber directly exposed to curing water in Chandler Engineering autoclaves, and outside of the curing chamber in direct contact with the chamber external wall for Jiangsu Hongbo autoclaves. The different design methodology lead to different temperature and pressure profiles during the ramp-up and ramp-down period (See Figure 6), where temperature was measured inside the curing chamber in both cases. Additionally, the pressure of Jiangsu Hongbo autoclaves was controlled by fully automatic syringe pumps, which allows much more precise control during both curing and ramping periods. Slurry Z1 and Z2 was cured in the two different types of autoclave under the condition of 200°C and 50 MPa for a period of 14 d. Test results are shown in Figure 7. There was almost no difference in the water permeability test results for specimens produced by different autoclaves. However, the compressive strength and Young's modulus of specimens produced by the Jiangsu Hongbo autoclave are significantly higher than those produced by the Chandler autoclave. These results suggest that the curing chamber design has a strong influence on the experimental results, possibly due to the different temperature and pressure profiles. As discussed earlier, Reddy et al. [12] had previously reported that heating rate during initial temperature ramp had a strong influence on the sonic strength development of oil well cement under HPHT conditions.

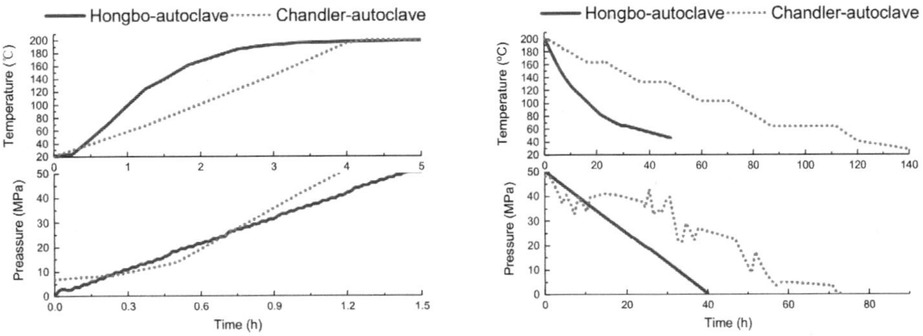

Figure 6. Temperature and pressure profiles during ramp-up and ramp-down of different autoclaves

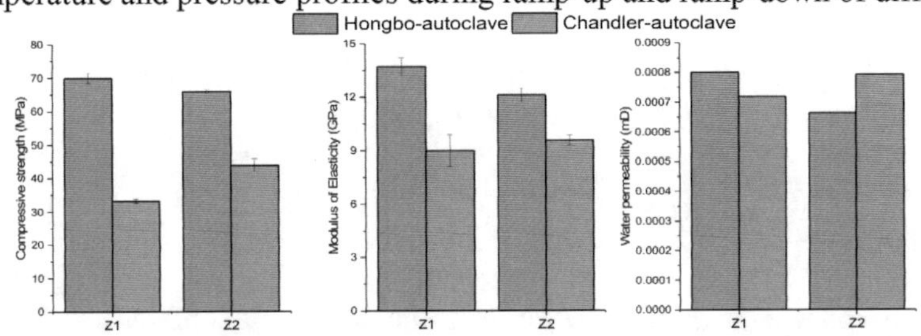

Figure 7. Influence of curing autoclave on experimental results (200°C and 50MPa, 14d)

4. Conclusions

Five influence factors on experimental results of oil well cement cured under high temperature and high pressure condition were studied in this article. The main conclusions are as follows:

(1) The long-term strength retrogression of oil well cement is dependent on the formulation design. Slurry Z2 of this study is much more stable than the other formulations during the curing period from 14d to 69d.

(2) Quick depressurization after high pressure curing will lead to specimen damage, which can significantly increase its water permeability; decreases in mechanical property is also observed in certain formulation. Slow depressurization method is preferred to improve the quality of the samples.

(3) Increasing curing pressure (from 50MPa to 150MPa) had little effect on the permeability and mechanical property test results of the formulation without silica fume, but improved the properties of the formulation with silica fume.

(4) For the same depressurization rate, bigger-size specimens are much more susceptible to damage than smaller-size specimens. The specimen damage can lead to significant deterioration in macroscopic mechanical properties, but no influence on microscopic mechanical properties.

(5) The curing chamber design can have a significant influence on the experimental results possibly because of the different temperature and pressure profiles. Therefore, it is critical to use the same type of curing chamber for the same test series.

Acknowledgement

Financial support provided by PetroChina Tarim Oilfield Company and China University of Petroleum (East China) (No. 2018000025) is greatly appreciated.

References

[1] X. Pang 2011 Effect of curing temperature and pressure on the chemical, physical, and mechanical properties of Portland cements. Ph.D. dissertation, Columbia University
[2] K.J. Krakowiak, J.J. Thomas, S. Muss, S. James, A.T. Akono, F.J. Ulm 2015 Nano-chemo-

mechanical signature of conventional oil-well cement systems: effects of elevated temperature and curing time *Cement and concrete research* **67** 103-121

[3] R.B. Pernites, A.K. Santra 2016 Portland cement solutions for ultra-high temperature wellbore applications *Cement & concrete composites* **72** 89-103

[4] O. Omosebi, H. Maheshwari, R. Ahmed, S. Shah, S. Osisanya, S. Hassani, G. DeBruijn, W. Cornell, D. Simon 2016 Degradation of well cement in HPHT acidic environment: effects of CO2 concentration and pressure *Cement & concrete composites* **74** 54-70

[5] R.W. Pollock, W.H. Beecroft, L.G. Carter 1966 Cementing practices for thermal wells *Journal of canadian petroleum technology* **5(3)** 130-134

[6] X Pang, C Meyer, GP Funkhouser, R Darbe 2015 Depressurization damage of oil well cement cured for 3 days at various pressures *Construction and building materials* **74** 268-277

[7] A.A. Mahmoud, S. Elkatatny, A. Ahmed, M. Mahmoud 2018 Nanoclay content influence on cement strength for oil wells subjected to cyclic steam injection and high-temperature conditions

[8] B. Iverson, J. Maxson, D. Bour 2010 strength retrogression in cements under high-temperature conditions, 35th workshop on geothermal reservoir engineering 2010, Stanford, California, USA 111-118

[9] B Iverson, B Waugh, J Maxson 2014 High-temperature exposure of oil well cements *Characterization of minerals, metals, and materials* San Diego, California, USA 17-24

[10] L. Li, M. Kellum, A. Doan 2016 In-situ tensile strength testing: awareness of variations with testing environment, spe deepwater drilling & completions conference

[11] W.C. Jimenez, X. Pang, M. Benge, R. Darbe, D.L. Meadows 2017 Enhanced mechanical integrity characterization of oilwell annular sealants under in-situ downhole conditions, spe oil & gas india conference & exhibition

[12] B.R. Reddy, J. Zhang, M. Ellis 2016 Cement strength retrogression issues in offshore deep water applications - do we know enough for safe cementing

[13] A. Anya, H. Emadi, M. Watson 2019 An empirical model for calculating uniaxial compressive strength of oil well cements from ultrasonic pulse transit time measurements, *J. Pet. Sci. Eng.* **183** 12

[14] M.A. Swayze 1954 Effect of high temperatures on strength of oil-well cements (mid-continent district study commitee on cementing practices and testing of oil-well cements), *Vacuum* **13(7-8)** 206

[15] X. Pang, D.P. Bentz, C. Meyer, G.P. Funkhouser, R. Darbe 2013 A comparison study of Portland cement hydration kinetics as measured by chemical shrinkage and isothermal calorimetry *Cement & Concrete Composites* **39** 23-32

[16] X. Pang, W.C. Jimenez, B.J. Iverson 2013 Hydration kinetics modeling of the effect of curing temperature and pressure on the heat evolution of oil well cement, *Cement and Concrete Research* **54** 69-76

[17] X. Pang, C. Meyer, R. Darbe, G.P. Funkhouser 2013 Modeling the effect of curing temperature and pressure on cement hydration kinetics, *ACI materials journal* **110(2)** 137-147

[18] X. Pang, P.A. Otieno, G.P. Funkhouser 2014 Modeling the effect of curing pressure on the viscosity evolution of oilwell cement, the 2014 aade fluid technical conference & exhibition

[19] G.W. Scherer, G.P. Funkhouser, S. Peethamparan 2010 Effect of pressure on early hydration of class H and white cement, *Cement and Concrete Research* **40(6)** 845-850

[20] A.C. Jupe, A.P. Wilkinson, K. Luke, G.P. Funkhouser 2008 Class H cement hydration at 180 degrees C and high pressure in the presence of added silica, *Cement and Concrete Research* **38(5)** 660-666

[21] C. Zhou, F. Ren, Z. Wang, W. Chen, W. Wang 2017 Why permeability to water is anomalously lower than that to many other fluids for cement-based material, *Cement and Concrete Research* **100** 373-384

IOP Conf. Series: Materials Science and Engineering 1028 (2021) 012005 doi:10.1088/1757-899X/1028/1/012005

The effect of the thickness of steel pipe in linear and 2-dimensional stiffness analysis during design of transmission tower

Yuxi Song*, Xianqi Chen, Zijian Di and Yihang Sun

School of mechanical and power engineering, east China university of science and technology, P.R. China, 200237

*Email: song_yvxi@sohu.com

Abstract. Transmission towers are truss structures which are used to support the transmission power lines. It is important to design the transmission towers because unsuitable dimensions could waste the steel or lead to the failure of transmission towers. The thickness of the beam is a critical parameter during the design of transmission tower, which is directly related with the safety of the transmission tower. In this paper, a finite element method in linear and 2-dimension was employed to simulate the mechanical behaviour of transmission tower. Based on the calculated results, the variations of the maximum stress and the maximum displacement with the increasing thickness were displayed. Both the stress and displacement decrease dramatically with the increasing thickness and reach to a stable value if the thickness is beyond a critical thickness. When the thickness changes from 1 mm to 4 mm, the maximum stress and the maximum displacement decrease by 74%. When the thickness is more than 4 mm, the reduction of stress and displacement is not obvious. The percentage reduction of more than 70% is determined to be the criterion for optimum thickness. The relationships can be used in the design of the transmission tower.

1. Introduction

The mechanical properties of transmission towers have been researched from last century. Ultimate loading is an important parameter related with the safety of transmission tower [1]. The ultimate loading can be obtained by measuring the stress of transmission tower during experiments [2, 3]. The periodic for the design of transmission tower could be longer if the experiment is only used. With the development of finite element software, the ultimate load and mechanical responses can be predicted by finite element analysis [4-6]. Based on the numerical results, the design of transmission towers can be modified to improve the safety [7]. The effect of the geometry on the strength of transmission towers was studied by a large scale simulation [8]. Numerical results revealed the limitation of the height of transmission towers [9]. The stress distribution and initial stress were calculated by finite element model [10, 11]. Recently, the numerical models for transmission tower could be quickly carried out by Solidworks, Workbench ANSYS and ABAQUS [12, 13]. Moreover, based on the numerical results in finite element software, Borja et al. optimized the transmission tower by the stress distribution within the transmission tower [14]. The previous works concentrate on the effect of height, loading force, stiffness and wind loading on the mechanical responses of transmission tower. Though the beam thickness of transmission power is related with the stiffness, the sole effect of beam thickness on the mechanical responses of transmission tower has not been investigated. The thickness

IOP Conf. Series: Materials Science and Engineering 1028 (2021) 012005 doi:10.1088/1757-899X/1028/1/012005

of beam is a critical parameter during the design of transmission tower, which is directly related with the safety of the transmission tower. The optimum thickness can save the usage of steel and ensure the enough stiffness; therefore, it is needed to research the mechanical responses of transmission tower at a range of beam thickness. Here, commercial finite element software (ABAQUS 6.14-4) was employed to analyze this variation. The relationships between the thickness and mechanical response were given out for the design of the transmission tower.

2. Finite element model

2.1. Geometric model

Power transmission tower is mainly composed of truss structures, and also the truss structure is the major stress component of power transmission tower. The power transmission tower model was built directly in the Abaqus, which is 100m in height, 20m in width at the bottom, 8m in width at the top, and with three pairs of wings as shown in Figure 1.The whole truss structure consists of pipe with radius of 0.05m and thickness varying from 1 mm to 10 mm. The truss is connected to the ground by two hinged joints on the bottom of the truss. Then the model is imported into the finite element program and is forced diagonally to the right with a load, which is a concentrated force of 10000N.

2.2. Material model

In the engineering application, the deformation of transmission tower is not allowed to be plastic deformation. Furthermore, the safety factor is considered to make sure the safety of transmission tower. The safety factor is commonly more than one; so that the allowable stress is smaller than the yield strength. When the transmission tower is designed, the stress value is calculated in its elastic range. Therefore, the analysis is linear, 2-dimensional and no failure criteria. The main diagonal material and auxiliary materials of the tower are made of steel with the elasticity modulus of 210GPa, a Poisson's ratio of 0.3 and density of 7800kg/m^3.

2.3. Numerical model

The mesh structure is divided in to seed parts with an approximate global size of 0.2m. There are 1320 nodes and 1400 linear line elements of type B31 in total. The finite element model for the transmission tower was shown in Figure 2, where the loading force and the boundary conditions were illustrated. The simulation conditions are displayed in Table 1. The thickness of pipe varies from 1 mm to 10 mm and ten simulated cases were carried out here. Because the deformations of tension, compression and bending can be happen, the Mises equivalent stress was chosen to reflect the mechanical response of transmission tower.

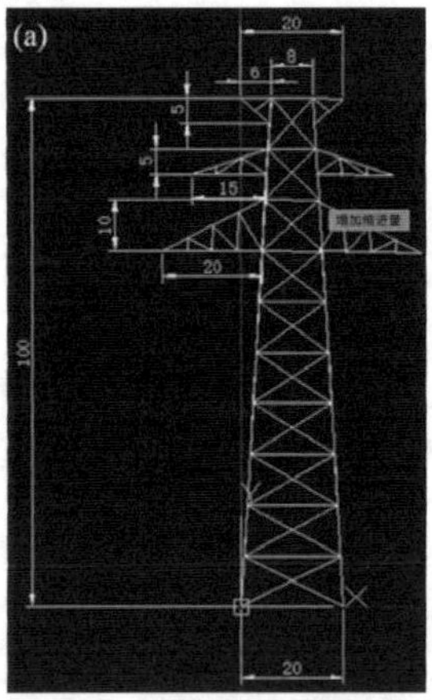

Figure 1. The dimensions for the transmission tower.

Figure 2. The finite element model for the transmission tower.

Table 1. Structural parameters of pipe for transmission tower.

No.	1	2	3	4	5	6	7	8	9	10
Diameter (cm)	10	10	10	10	10	10	10	10	10	10
Thickness (mm)	1	2	3	4	5	6	7	8	9	10

3. Mechanical response

In engineering practice, it is necessary to study the stress of power transmission tower, that is subjected to different types of forces as wind, snow, rain and so on, and whether the induced stress is relatively well-distributed or not, namely the maximum stress and the distribution of the stress and the displacement of the power transmission tower. Figure 3 shows the stress distribution for the pipe thickness of 5mm. The maximum stress value of power transmission tower is 32.7 MPa which is occurred at the bottom of communication tower. The maximum stress concentrates on the parts where the tower is hinged to the ground. And the stress is maximum at the bottom and becomes smaller at the top part of the tower. On the contrary, as shown in Figure 4, the displacement is smaller when the position is closer to the hinged joints and the maximum appears on the top of the transmission tower. For the pipe thickness of 5mm, the maximum displacement is about 82.3 mm.

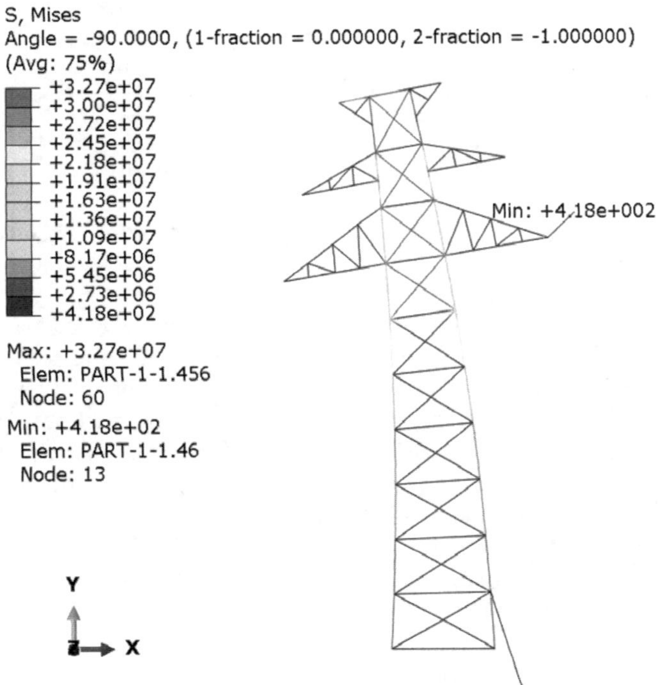

Figure 3. The stress distribution in transmission tower at the pipe thickness of 5mm.

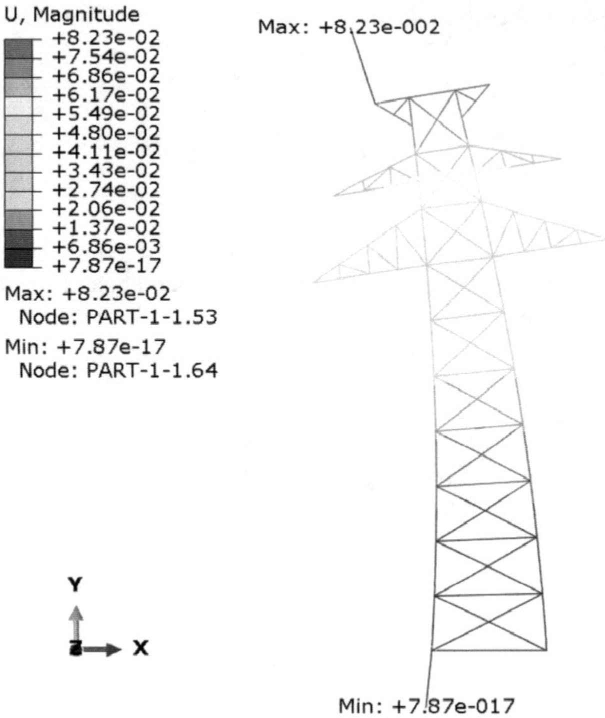

Figure 4. The stress distribution in transmission tower at the pipe. thickness of 5mm.

By increasing the thickness of the pipe from 1 mm to 10 mm, the relationship between the thickness and the mechanical responses was obtained. Based on the calculated data, the variations of maximum stress and maximum displacement with the increasing thickness are illustrated in Figure 5. Both the maximum values of the stress and the displacement decrease with the increasing pipe thickness. If the pipe thickness varies from 1 mm to 4mm, the maximum stress and the maximum displacement dramatically decrease. The maximum stress and the maximum displacement vary slightly with the increasing thickness when the pipe thickness is beyond 4mm. That is to say, when the pipe thickness is larger than a critical value, there is no need to further increase the pipe thickness. The larger thickness could lead to the waste of steel materials. Based on the standard of DL/T5254-2010, the optimum thickness is 4mm. When the thickness changes from 1 mm to 3 mm, the maximum stress and the maximum displacement decrease by 66%. When the thickness changes from 1 mm to 4 mm, the maximum stress and the maximum displacement decrease by 74%. When the thickness changes from 1 mm to 5 mm, the maximum stress and the maximum displacement decrease by 79%. According to the optimum thickness of 4 mm in DL/T5254-2010, the percentage reduction of more than 70% is chosen to be criterion for optimum thickness. From the calculated results in Figure 5, the optimal pipe thickness is 4 mm for the design of transmission tower whose dimension parameters are shown in Figure 1.

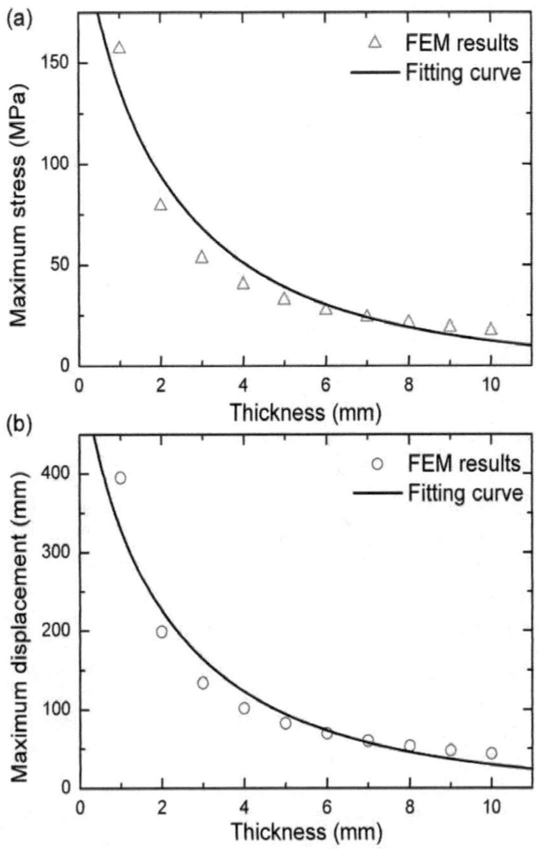

Figure 5. Variations of maximum stress and displacement with the increasing thickness (a)Variations of maximum stress with the increasing thickness; (b)Variations of maximum displacement with the increasing thickness.

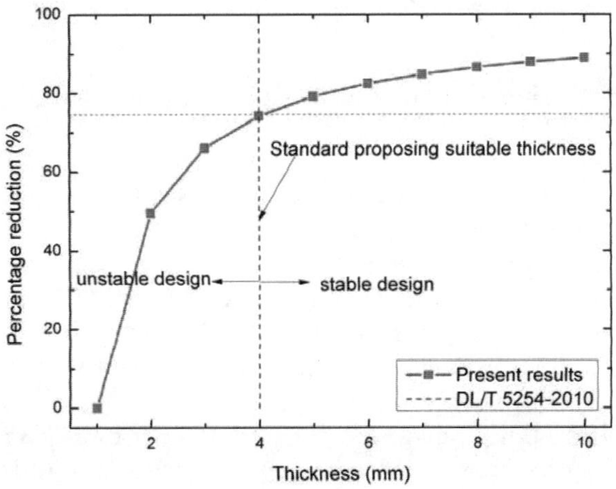

Figure 6. Comparison between the present results and the available standard.

Based on the calculated results, the relationships between mechanical responses and wall thickness were fitted by means of exponential function. Eq. (1) shows the exponential relationship between maximum stress and wall thickness:

$$\sigma_{max} = 250\exp(-0.6t^{0.7}), \tag{1}$$

where σ_{max} is the maximum stress and t is the wall thickness.

The fitting function in Figure 5b is in the following form:

$$S_{max} = 600\exp(-0.6t^{0.7}), \tag{2}$$

where S_{max} is the maximum displacement.

According to Eqs. (1)-(2), the relationships between maximum stress, maximum displacement and wall thickness are exponential. The coefficients in exponential function are different from each other.

Figure 6 shows the comparison between the present results and available design standard. In DL/T 5254-2010, the proposed thickness is 4 mm. If the thickness is less than 4 mm, the designed transmission tower will be in a risk of collapse. Here, we reveal the underlying mechanism on the optimum thickness by conducting a series of numerical calculations. It is illustrated in Figure 6 that the percentage reduction will increase gradually with the increasing thickness if the thickness is beyond 4 mm. The percentage reduction of more than 70% is chosen to be criterion for optimum thickness. The criterion of more than 70% percentage reduction can be used more widely than the available standard because the standard is employed in the special conditions. Moreover, the criterion of more than 70% percentage reduction is more suitable to be implemented into finite element software with the rapid development of commercial finite element method.

4. Concluding remarks

In this paper, a finite element model for the transmission tower was established by means of ABAQUS 6.14-4 software. The pipe thickness varied from 1 mm to 10 mm in the proposed finite element model. Based on the calculated results, the following conclusions are drawn:

(1) The relationship between the mechanical responses and the pipe thickness is nonlinear.

(2) With the increasing pipe thickness, both the maximum stress and the maximum displacement decrease dramatical for small thickness and then decrease slightly for large thickness.

(3) There is an optimal pipe thickness beyond which the thickness has a slight effect on the mechanical responses of the transmission tower.

(4) The percentage reduction of more than 70% is chosen to be criterion for optimum thickness. The optimum thickness for transmission tower is 4 mm by employing the proposed criterion.

Acknowledgement

This research was financially supported by Fundamental Research Funds for undergraduates.

References

[1] SL Chan, SH Cho 2008 Second-order analysis and design of angle trusses: elastic analysis and design [J] *Eng struct* **30 (3)** 616–25

[2] Hong Zhou Deng, Bin Huang 2018 Study on ultimate bearing capacity of main member in transmission tubular tower leg [J] *Thin-walled structures* **127** 51-61

[3] Qianjin Shu, Zhaohui Huang, Guanglin Yuan, Weiqing Ma, Sheng Ye, Jing Zhou 2018 Impact of wind loads on the resistance capacity of the transmission tower subjected to ground surface deformation [J] *Thin-walled structures* **131** 619-630

[4] Xiaoqi Zhou, Lingfeng Cai, Jingtai Etc 2011 Compact narrow ANSYS modeling and dynamic analysis of transmission tower [J] *Water resources and power* **29** 167-168

[5] Meiguo Li, Yandong, Wangjin 2011 Finite element analysis of hard and soft combination tower structure [J] *Coal electrical machinery* **4** 86-87

[6] QJ Shu, GL Yuan, GL Guo, YF Zhang 2012 Limits to foundation displacement of an extra high voltage transmission tower in a mining subsidence area [J] *Int J min sci technol* **22** 13–18

[7] F Yang, Q Li, J Yang, B Zhu 2013. Assessment on the stress state and the maintenance schemes

of the transmission tower above goaf of coal mine [J] *Eng fail anal* **31** 236–247

[8] Y Zhuge, JE Mills, X Ma 2012 Modelling of steel lattice tower angle legs reinforced for increased load capacity [J] *Eng struct* **43** 160–168

[9] Prasad Rao, Samuel Knight, SJ Mohan, N Lakshmanan 2012 Studies on failure of transmission line towers in testing [J] *Eng struct* **35** 55–70

[10] C Klinger, M Mehdianpour, D Klingbeil, D Bettge, R Häcker, W Baer 2011 Failure analysis on collapsed towers of overhead electrical lines in the region Münsterland (Germany) 2005 [J] *Eng fail anal* **18** 1873–1883

[11] Dehe Wang, Hao Fu 2018 Study of the distribution of initial stress for mechanical properties of guyed transmission tower [J] *Advances in intelligent systems research* **143** 349-353

[12] XW Kang, YW Zhang 2016 Dynamic analysis of communication transmission tower based on finite element method [J] *International conference on artificial intelligence and engineering applications* **63 (5)** 75-79

[13] Prasad Rao, V Kalyanaraman 2001 Non-linear behaviour of lattice panel of angle towers [J] *J. Constr steel Res* **57** 1337–1357

[14] C Borja, V Alberto, C Manuel, G Diego 2015 Geometrical Issues on the Structural Analysis of Transmission Electricity Towers Thanks to Laser Scanning Technology and Finite Element Method [J] *Remote sensing* **7 (9)** 34-42

Collapse behavior simulation of cylindrical steel shell under landslide impact

Menghong Wang[1], Feiyu Shi[1,*], Yaoxiang Zhao[1] and Xiaobin Zheng[1]

[1]School of civil and transportation engineering, Beijing university of civil engineering and architecture, 100044, Beijing.

[*]Email:2108140218002@stu.bucea.edu.cn

Abstract. ANSYS/LS-DYNA dynamic analysis software was adopted to establish a model of a grid structure impacted by a landslide load, with the dynamic effects of impact loads and contact collision theory considered based on Hamilton's principle. The case analysis in this study can provide guidance for anti-slump design in similar projects.

1. Introduction

Space grids are widely used for public large-span space structures such as airports, stations, and stadiums. In recent years, owing to the environmental protection requirements, a large number of storage silo structures have been used in power plants, steel mills, and cement plants, with spans reaching 150 m or more. Because these large-span storage silos are often located in factory areas, structural collapse often occurs under the action of strong winds, heavy snow, or strong earthquakes which exceed the design specifications. Local collapse and total failure of these structures caused by landslides also occur occasionally.

The ANSYS/LS-DYNA dynamic analysis software is used to simulate the collapse behavior of grid structures under unconventional design loads such as explosions, missiles, landslides, and rockfalls. It is necessary to apply dynamic analysis, basic contact collision theory, and geometric and material nonlinearity comprehensively to predict the possible collapse of the structure more accurately. This study is focused on improving design collapse resistance through local strengthening to prevent problems proactively. Most previous studies [1-6] have focused on single point impacts on single-layer reticulated shells, but there are few studies simulating the effects of large-scale landslides.

This study investigates the partial collapse of a grid structure caused by a landslide, and a model of an actual grid is simulated during the impact of a landslide. Based on the energy change and dynamic response of the structure, the failure mode of the structure resulting from landslide damage is discussed in detail.

2. Basic theory of impact dynamics

2.1. Establishment and solution of the dynamic equilibrium equation

The analysis of structural dynamics requires establishment of structural dynamic equilibrium equations based on the structure of interest and the loads applied. A reasonable solution method must then be applied for the numerical calculation. The most common method for establishing the dynamic equilibrium equation uses D'Alembert's principle to balance the equation directly, but this method is not suitable for complex structures. For complex structures, the motion equation can be established using the virtual displacement principle, but only vector addition can be permitted. By applying Hamilton's principle to establish the equation of motion, the disadvantages of D'Alembert's principle and the virtual displacement principle can be avoided. Hamilton's principle is expressed as Eq. (1).

$$\delta J_H = \delta\int_{t_1}^{t_2} L dt = \delta\int_{t_1}^{t_2} (T - V) dt + \delta\int_{t_1}^{t_2} W_{nc} dt = 0 \tag{1}$$

where J_H is the Hamilton effect; L is the Lagrange function; T is the total kinetic energy, $T = m\dot{u}^2/2$; V is energy of position, $V = Ku^2/2$; W_{nc} is the work of non-conservative forces on the system; and δ is the variation within the specified time period.

Hamilton's principle can be described as follows: in any time period, the sum of the variation in the kinetic energy and energy of position plus the non-conservative force is always zero. The Hamilton principle differs from the virtual work method in that it uses variations in the kinetic energy and potential energy instead of the inertial force and elastic force. The equation established with this method is only related to the scalar energy, and is therefore more applicable to situations in which an impact load acts on the structure [7].

The explicit center difference method used in ANSYS/LS-DYNA represents the speed and acceleration in the equation of motion by different combinations of displacements, which converts the hard-to-solve differential equations into algebraic equations that can easily be solved numerically, and derives the recurrence formula for each time step. The structural reaction during the entire exercise can then be obtained. Thus, this study uses the explicit center difference method to solve the equation.

2.2. *Dynamic constitutive model of the materials*

Cowper-Symonds constitutive equation was employed to consider the effect of the material strain rate. The equation (2) expresses the relationship between the strain rate and the flow stress; the temperature factor is not included, and only the influence of the strain rate is considered [8].

$$\sigma_y = \left[1 + \left(\frac{\dot{\varepsilon}'}{C}\right)^{\frac{1}{P}}\right]\left(\sigma_0 + f_h\left(\varepsilon_{eff}^p\right)\right) \tag{2}$$

where σ_0 is the yield stress, $\dot{\varepsilon}'$ is the effective strain rate, C, P are strain rate parameters, and $f_h\left(\varepsilon_{eff}^p\right)$ is the hardening function for the effective plastic strain.

2.3. *Basic theory of contact and collision*

Common contact algorithms include the dynamic constraint method, distribution parameter method, and penalty function method. The penalty function method is used as the contact algorithm in this study. The basic principle of the penalty function method is to check the boundaries of each major surface for each Δt to determine if the boundary has been penetrated from the node. If not, processing is performed. Otherwise, a contact force is exerted on the interface, which can be calculated according to Eq. (3), and is equivalent to adding a series of springs between the interfaces.

$$F = k\delta \tag{3}$$

Where, k is the contact surface stiffness, and δ is the contact surface penetration.

3. Treatment of landslide impact

Simulation of an actual landslide is relatively complicated and requires consideration of many factors. Because the primary impact material in the landslide in this case is mountain gravel and the actual amount of earthwork is small, the treatment can be simplified as stones impacting the grid. When the grid is impacted by stones, the stones at the lower grid structure are larger and more numerous, while the stones at higher locations are lighter and less abundant. Therefore, the following assumptions are made:

1) The mass of the stones decreases linearly from the bottom to the top of the structure, with an equilateral triangle distribution;

2) The site of the impact by the stone occurs at the node of the grid;
3) The stones are not broken and are spherical in shape.

4. Grid finite element model

This study uses an actual structure as the analysis model, and this structure has a span of 50 m × 248 m, a height of 20.438 m, and an unfolded area of 19,880 m^2. The perimeter of the grid is fixed by a three-way hinged support.

Based on the symmetry of the grid structure, 1/2 of the overall structure is used for the simulation. The grid bars are modeled using beam element BEAM161. For the bars, a piecewise linear plastic material model was used, with a yield strength of 207 MPa, elastic modulus of 206 GPa, and Poisson's ratio of 0.3. The landslide impact material was simulated using the solid element SOLID164. To facilitate modification of the parameters, the impact material was defined as a sphere with a radius of 0.62 m and volume of 1 m^3. The roof dead load and bolt ball weight were simulated using MASS166.

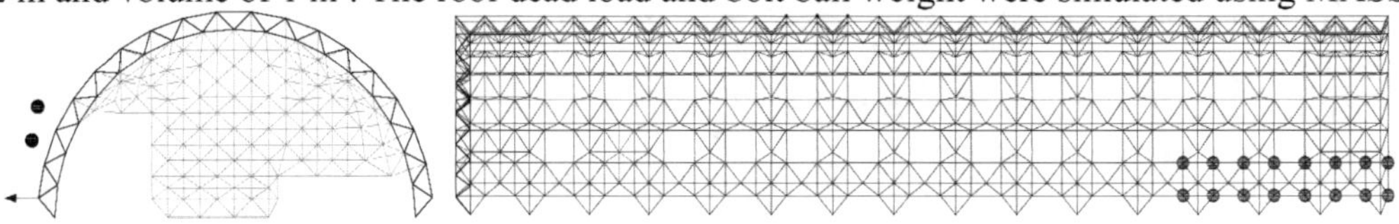

(a) Left view (b) Front view

Figure 1. Views of the structural model.

The initial position of the landslide impactor is in the positive y-direction from the impacted node. The change in weight of the landslide is achieved by varying the material density of the impact material, and the density of the lower impact material is greater than the density of the upper impact material. The landslide impact grid is realized by impacting the grid from a horizontal direction with a certain initial velocity. The model is shown in Figure 1.

5. Structural dynamic response

The failure modes can be divided into three situations: 1) slight damage leading to a dent (less than 0.5% of the bars in the impact area); 2) failure of a large number of bars in the impact area (less than 0.5% of bars in the impact area); and 3) overall structure failure with a large number of bar failures (the overall structure is greater than 0.5% of the bar failure).

The simulation results show that Mode 2 is more similar to the results of an actual landslide. A lower monomer mass of 1500 kg, upper monomer mass of 750 kg, and impact velocity of 50 m/s. The impact process is shown in Figure 2. After the simulated impact object collides with the grid, the nodes in the impact area move horizontally together. The neighboring nodes around the impact area also move horizontally along with the impact area.

When the displacement of the horizontal movement reaches a maximum, some of the impact objects rebound, and the grid will vibrate and consume energy to reach a certain equilibrium position. This process is accompanied by the failure of some of the bars. Eventually, a few bars in the impact area of the grid will fail.

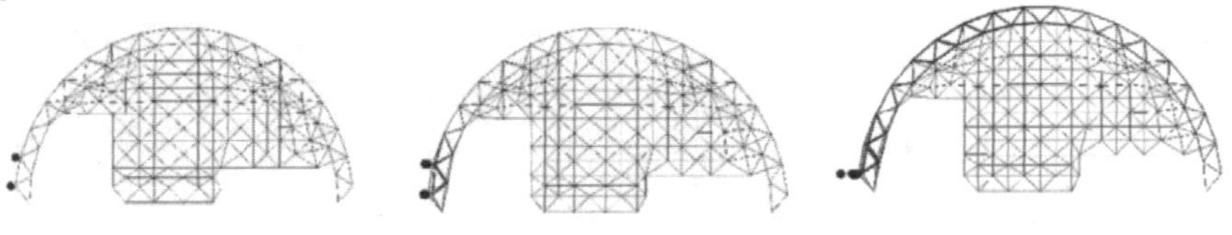

(a) t=0 s (b) t=0.017 s (c) t=0.5 s

Figure 2. Impact process.

5.1. Kinetic and strain energy of the grid

The kinetic energy history curve for the grid is shown in Figure 3. Immediately after the grid node is impacted, the kinetic energy of the grid increases.

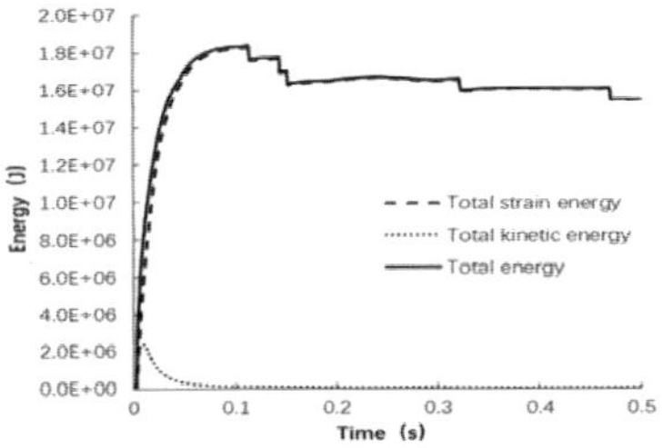

Figure 3. Grid energy time history.

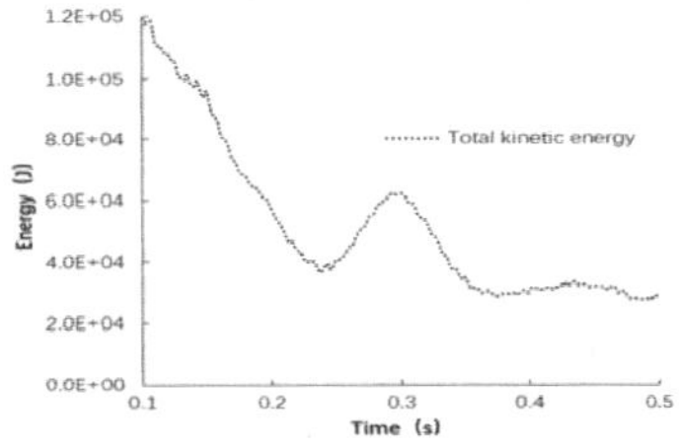

Figure 4. Time history of the total kinetic energy of the grid.

At 0.009 s, the kinetic energy reaches a maximum value of 2.4×10^6 J, after which the kinetic energy decreases rapidly, and the kinetic energy is quickly transformed into strain energy in the grid. After 0.2 s, the kinetic energy of the grid is relatively small, i.e., almost zero, indicating that the kinetic energy of the grid has been converted into strain energy.

At 0.1 s, the strain energy of the grid increases to 1.83×10^7 J. After 0.4 s, the strain energy of the grid begins to decreases stepwise. At 0.11 s, 0.14 s, 0.15 s, 0.32 s, and 0.469 s, the strain energy of the grid undergoes a sudden change. Finally, the strain energy of the grid stabilizes at 1.54×10^7 J. This is because at each of the above five moments, one bar member reaches the maximum plastic strain, which causes it to lose its ability for load bearing, and a portion of the energy is consumed.

The total kinetic energy time history of the grid is shown in Figure 4. The kinetic energy increases again at 0.24 s. This is because the impact area of the grid begins to rebound after the horizontal displacement reaches its maximum value, and the grid vibrates around the equilibrium position.

5.2. Node y-direction velocity

The lower layer is impacted at node 3454, and the upper layer is impacted at node 3453. Nodes 3456 and 3458 are laterally adjacent to the impact point. When the impact object contacts the grid, the velocity of the impacted node increases sharply. The velocity of node 3454 reaches -52.0 m/s within 1 ms, while node 3453 reaches -18.1 m/s. As the impact loading ends, the velocity at the impact point no longer increases. Furthermore, the speed decreases rapidly due to the need to drive the horizontal movement of neighboring nodes. The duration of the impact is very short. At this time, the velocity development at adjacent nodes 3456 and 3458 is delayed, and the fluctuation is not severe. Driven by nodes 3454 and 3453, the velocities of adjacent nodes 3456 and 3458 increase continuously, and the y-direction velocities of the nodes all gradually decrease toward zero. The node velocity variation at each corresponding position is basically the same as the y-direction velocity variation observed for the impacted node and adjacent nodes in the third column.

5.3. Node y-direction displacement

The y-direction displacements for the impacted nodes and adjacent nodes in the third column are shown in Figure 5(a). At the moment of impact, the displacement at the impacted point increases sharply. Within 60 ms, the y-direction displacement of node 3454 reaches -0.902 m, and the displacement of node 3454 reaches -0.779 m. Because the impact duration is extremely short, the displacements of adjacent nodes 3456 and 3458 are delayed. Driven by nodes 3454 and 3453, the displacements of adjacent nodes 3456 and 3458 continuously increase, and the displacements of the nodes in the y-direction then all gradually decrease and tend to stabilize. During the entire impact process, the time of maximum displacement increases sequentially from the bottom up to the four nodes in turn, which is consistent with the change in an actual impact.

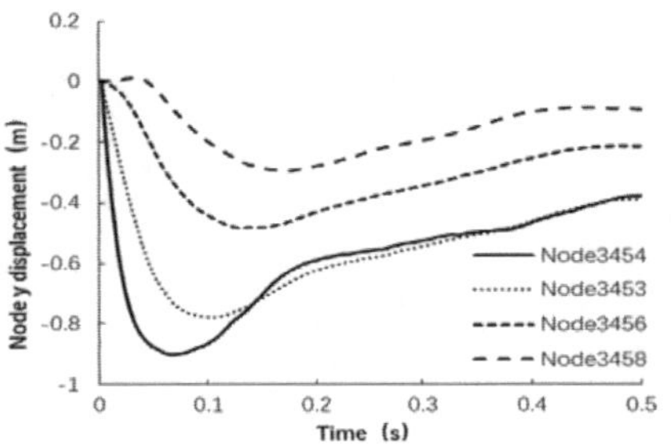

(a) y-direction displacements for impact at the third column

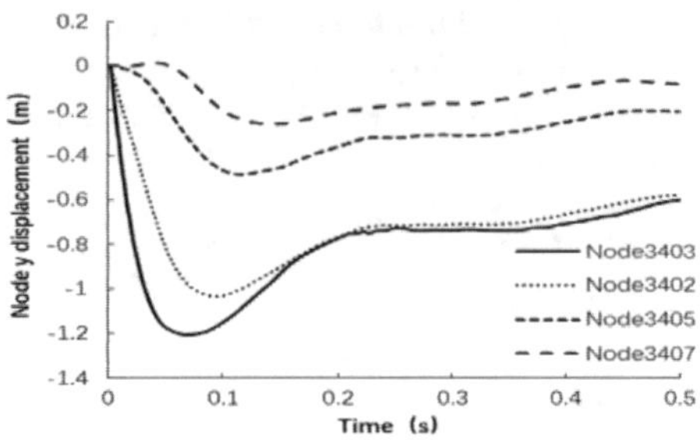

(b) y-direction displacements for impact at the sixth column

Figure 5. Y-direction displacements of impacted and adjacent nodes.

The y-direction displacements for the impacted nodes and adjacent nodes in the sixth column are shown in Figure 5(b). The change in the node displacement at each corresponding position is basically the same as the y-direction displacements observed for the impact and adjacent nodes in the third column.

5.4. Stress of the bars

The stress time history curve for bars at the impact position in the third column and an adjacent bar are shown in Figure 6(a). The compressive stress of bar 2410 is 622 MPa at 0.053 s. The tensile stress of bar 2411 is 267 MPa at 0.015 s. The tensile stress of bar 2412 is 366 MPa at 0.031 s. The tensile stress of bar 2413 is 345 MPa at 0.041 s. The instantaneous stresses of these bars are much greater than the yield strength of the steel under a static load, indicating that the strain rate effect has a significant influence on the structure of the grid under an impact load. The bar will continue to oscillate after the impact force ends.

The stress time history curve for bars at the impacted position in the sixth column and adjacent bars is shown in Figure 6(b). The stress time history curve of the axial stress for the bars in each corresponding position is basically the same as that observed for the impact position in the third column.

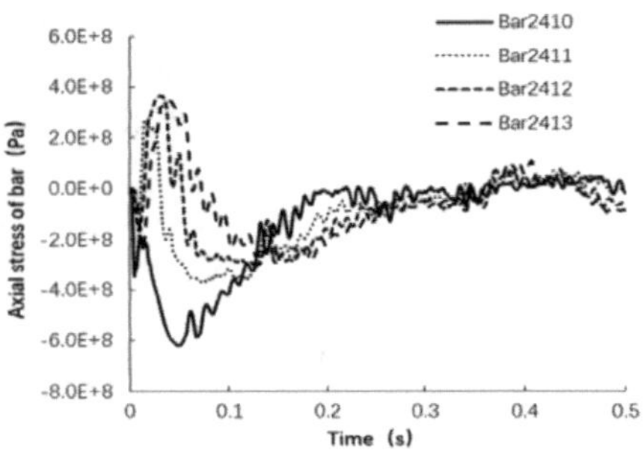

(a) Time history curve for the impact position in the third column

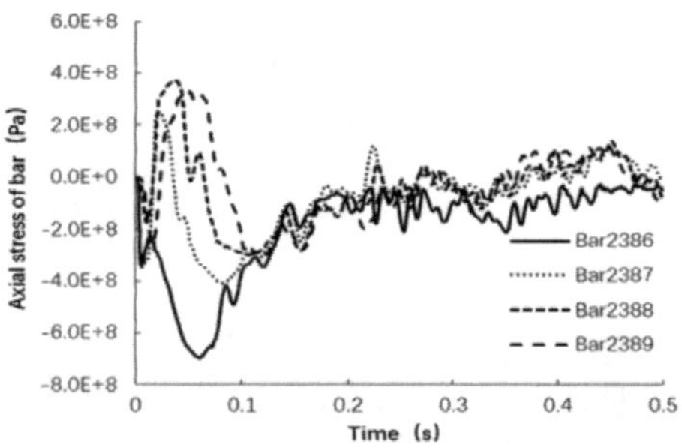

(b) Time history curve for the impact position in the sixth column

Figure 6. Bar stress time history curves.

5.5. Plastic development and plastic area of the grid structure

For the development of plastic strain in the impact process, Figure 7(a-c) shows that with the development of the impact process, most of the bars around the impact area enter plasticity. In this simulation, only a few bars failed. Most of the other bars entered the plastic stage. A bar entering plastic strain is deemed a damaged bar.

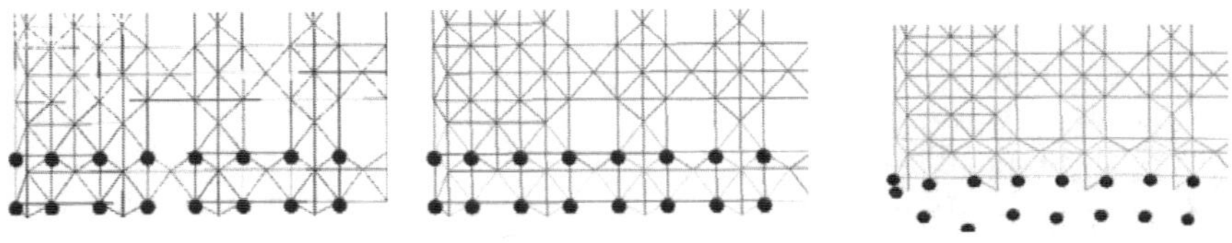

(a) t=0 s (b) t=0.0058 s (c) t=0.5 s

Figure 7. Plastic strain development during impact.

The distribution of bars entering plasticity is shown in Figure 8. Red bars indicate the occurrence of plastic strain. Most of the bars in the impact area enter the plastic stage. There are also a few bars in the non-impact area which enter the plastic stage. The plastic bars in the non-impact area mostly occur near the transverse upper and lower chords.

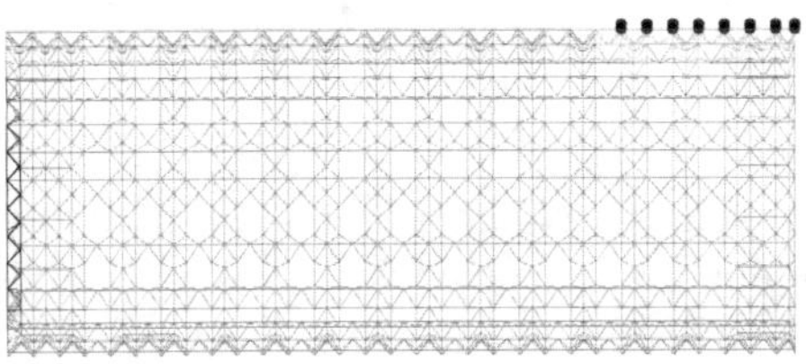

Figure 8. Distribution of plastic bars (shown in red).

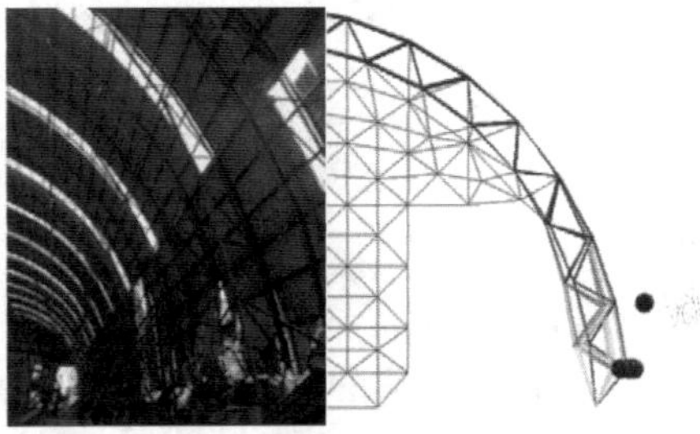

Figure 9. Comparison of actual grid damage and simulation results.

A comparison of the actual grid damage with the simulation results is shown in Figure 9. It can be seen that the simulation results in this study are roughly the same as the damage to the grid after the actual landslide impact.

5.6. Strengthening of key bars

Based on the above analysis, the key bars in the grid under an impact load are the bars that enter the plastic stage in this mode, i.e., the red bars in Figure 8. This study replaces the plastic bar members with Q345 members to increase their yield strength and thus achieve the effect of grid strengthening [9].

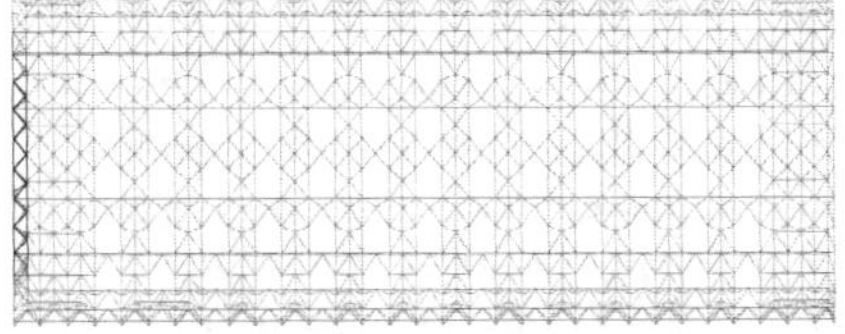

Figure 10. Distribution of reinforced plastic bars (shown in red).

The relevant analysis of the dynamic response discussed above is performed on the grids after strengthening the identified key bars. The same impact load is applied in the analysis. The resulting distribution of grid bars entering the plastic stage is shown in Figure 10. In the impact area, some of the bars entered plasticity, while only a very small number of bars in the non-impact area entered the plastic stage. After strengthening, the number of bars entering the plastic stage has been greatly reduced, and the failure mode has changed from Mode 2 to Mode 1.

The dynamic response of the strengthened grid under different impact masses and velocities was analyzed by changing the mass and velocity parameters of the impact material. After reinforcement, the failure mode of the grid is improved compared with the ability of the unreinforced grid to resist collapse.

After strengthening the key bars, under the same impact load, the number of bars entering the plastic stage is reduced significantly, and the structure is more resistant to collapse. The results also demonstrate that the reinforced design for key bars can enhance the collapse resistance of the structure under an impact load.

6. Conclusions

1) With increasing impact energy, the sequential failure modes of the double-layer cylindrical lattice are slight damage to the depression, failure of a large number of bars in the impact area, and failure of a large number of bars in the overall structure. The simulation results show that Mode 2 is consistent with the impact on a grid by an actual landslide.

2) Most of the bars around the impact area enter plasticity. The distribution of plastic bars shows that the plastic development of the structure does not extend to the entire grid. The grid bars in the non-impact area which enter plasticity are mostly located in the transverse upper and lower chords.

3) When the grid is impacted by a landslide, the impact area is impacted and gradually expands. The kinetic energy is then absorbed by the grid and decreases. Finally, the kinetic energy is almost completely converted into strain energy. The strain energy continues to increase and eventually stabilizes.

4) The stress of the bars under impact loading can be much greater than the yield stress under static loading. The steel material is thus a strain rate sensitive material. Whether or not the strain rate effect is considered in the structural analysis under impact loading can affect the results.

5) The key bars of the double-layer grid structure are determined under the impact load. By strengthening the design of the key bars, the collapse resistance of the structure under impact loading can be effectively enhanced. In addition, this method for improving the ability of the structure to resist collapse can also be used as guidance for related projects.

Acknowledgement

This study is sponsored by the BUCEA Post Graduate Innovation Project(PG2020029).

References

[1] Guo Ke 2004 Dynamic analysis of single-layer reticulated shells under impact [D]. Taiyuan: Taiyuan University of Technology
[2] Wang Duozhi 2010 Failure mechanism of reticulated shells under impact [D]. Harbin: Harbin Institute of Technology
[3] Fan Feng, Wang Duozhi, Zhi Xudong, et al 2009 Performance for kiewite8 single-layer reticulated domes subjected to impact load *Engineering mechanics* **26 (06)**75-81
[4] Fan Feng, Wang Duozhi, Zhi Xudong, et al 2010 Failure modes and discrimination method for Kiewitt reticulated dome under impact loads *China civil engineering journal* **43 (05)**56-62.
[5] Shi Junliang 2005 Dynamic response research on K8 single-layer reticulated dome under impact [D]. Taiyuan: Taiyuan University of Technology
[6] Tian Yonghong 2010 Failure modes of geodesic and schwedler reticulated domes under impact loads [D]. Harbin Institute of Technology
[7] Ray Clough, Joesph Penzien, Wang Guangyuan, et al. Dynamics of structures (Revised Edition) [M]. Beijing: Higher education press **2006** 7
[8] LSTC. LS-DYNA 2015 keyword user's manual[M]. California: Livermore Software Technology Corporation
[9] Xiong Mingxiang 2005 Study on impact response and protective measures of steel structure with composite slab under impact loads [D]. Wuhan: Huazhong University of Science and Technology

Research and application of anticorrosive concrete for shafts in thick and deep surface soil

Guisheng Cai[1,2], Jie Zhang[1,2,*], Tongwei Lu[1,2], Lei Guo[1,2] and Jinhuan Lv[1,2]

[1] Shandong academy of building research Co., Ltd, Ji'nan, China;
[2] Shandong jianke building materials Co., Ltd, Ji'nan, China.

[*]Email: jessiez1212@126.com

Abstract. Most frozen shafts of deep surface layer are located at a harsh environment, surrounding soil or groundwater with lots of corrosive ions such as SO^{4-} and Cl^-. This is deadly to the durability of concrete. Thus, concrete for shafts in deep surface soil requires excellent durability. This paper focuses on the development and preparation of anticorrosive concrete in deep surface layer shafts, and presents engineering application. The experimental results show that NC-H admixture is successfully made to prepare the concrete. The durability and impermeability of concrete are good. The Cl^- diffusion coefficient of C60 is $38.1\times10^{-14}m^2/s$, while C90 is only $4.5\times10^{-14}m^2/s$. Besides, high freezing and thawing resistance is discovered. The loss of dynamic elastic modulus of C75 is less than 6.38% when the freeze-thaw cycle is 1000 times, It has high anti-sulfate ability and no corrosion to steel bar. This anticorrosive concrete has applied in many large coal mining projects. It is proved that this anticorrosive concrete has been applied to practical engineering application, and it can completely meet the requirements of cast-in-situ reinforced concrete shaft construction with deep surface layer and corrosion resistance.

1. Introduction

As the more accessible portions of coal resources near the surface are gradually depleted in large coal. It is necessary to explore deeper coal seams, at present, most shafts traverse the topsoil layer during the construction process, and some shafts have topsoil layer thickness deeper than 500m [1-3]. In general, freezing method is used in construction of deep surface layer, due to its strong adaptability, high reliability and well completion speed [4]. Reinforced concrete shaft wall structure is commonly applied in freezing method [5, 6].

With increase of shaft depth, the thickness of concrete in shaft wall increases greatly. On the one hand, increased thickness lead to dead weight and high construction cost of concrete. On the other hand, it enlarges the volume of concrete, which easily causes cracks and not conducive to the safety and service life of projects [7]. Considering this, high strength and performance concrete were applied in shafts as it can reduce self-weight of concrete structure and reduce the material. It is an effective approach to solve the problem of freezing shaft wall design and construction in deep topsoil layer. Currently, C30-C40 and other low-grade concrete has been gradually replaced [2, 7, 8].

In recent years, lots of concrete shaft walls have been damaged to varying degrees, which requires a large amount of funds to be spent on maintenance and reinforcement [9-12].

This seriously affects safety of mine production. In addition to the strength of concrete itself, corrosion resistance and durability are significant challenges for concrete structures of concrete shaft walls in deep topsoil [13]. The deep surface soil, naturally, requires that shaft wall should bear various pressures, including huge ground pressure and vertical additional forces during the use of the shaft. Besides, it should bear the freezing pressure during the construction of the well in freezing method construction [1,6].

Moreover, deep topsoil owns high water content, pressure and water head. Thus, there exist water seepage, water leakage and corrosion of harmful ions (SO_4^{2-}, CO_3^{2-}, Cl^-, Na^+, K^+ and so on) in underground water. Some projects are located in in saline soil area with sulphate and part of chlorine salt which has a strong corrosion effect on concrete and steel reinforcement. Chlorine salt corrodes steel reinforcement and easily causes cracking of concrete structure, which destroys concrete structures [1,3]. Sulphate exposure is one of the most frequent and detrimental processes for the durability of concrete structures. The sulphate ions found in ground waters may diffuse through concrete pores and react with its components leading to the formation of expansive products which cause an expansion of the concrete leading to a decrease in the durability and more corrosion to reinforced bars [14].

There are many ways to improve the corrosion resistance of concrete. The general technical approaches are not only to improve corrosion resistance of concrete itself, but also to coat the surface of concrete with anti-corrosion coating [15]. However, for rust resistance of steel bars, the rust resistance layer should be painted on the surface of steel bars and the rust inhibitor should be added in concrete. These commonly used approaches for new concrete wellbore project is very difficult and not feasible. Therefore, improving corrosion resistance of concrete is more sensible and meaningful.

In this paper, C60~C90 anticorrosive concrete in deep surface layer shafts are developed and prepared successfully, and the properties are furthermore studied. It has been applied in some big projects, such as Juye coal field, Longgu coal mine and so on.

2. Experimental work

2.1 Materials

Cement include two types of Ordinary Portland cement 42.5 and 52.5, and the reference cement. Mechanical properties are given in Table 1.

Table 1. Mechanical properties of P.O 42.5 and P.O 52.5.

Cement grade	Fluidity(mm)	Flexural strength (MPa)			Compressive strength (MPa)		
		3d	7d	28d	3d	7d	28d
P.O 42.5	180	6.0	7.4	8.4	28.3	38.6	51.5
P.O 52.5	185	7.0	8.1	10.5	32.4	43.3	56.7
Reference	180	4.7	6.2	8.8	19.2	32.3	48.6

Table 2. Performance indicators of NC-H admixture.

Labels	Moisture content, ≤ (%)	Fineness, Over 0.045 mm sieve, ≤ (%)	Fluidity of cement paste≥ (mm)	Compressive strength ratio of mortar			Corrosion of steel reinforcement
				3d	7d	28d	
NC-H60	3.0	10.0	230	140	130	120	none
NC-H70	3.0	10.0	230	140	130	120	none
NC-H80	3.0	10.0	230	150	140	130	none
NC-H90	3.0	10.0	230	160	150	140	none

NC-H admixture developed by our previous researches, with characteristics of reducing water, expanding slightly and improving strength, is used. It mainly consists of mineral admixtures, chemical admixtures, micro expansion components, accelerating component and modifying agents. According to different concrete grades, NC-H is divided into a variety including NC-H60, NC-H70, NC-H80 and

NC-H90. The performance indexes of NC-H are shown in Table 2. For cement paste experiment, the content of NC-H60, NC-H70, NC-H80 and NC-H90 is respectively 10%, 10%, 15% and 15%. For motor, the content is 25%, 30%, 35% and 40%.

Fine aggregate is sand from river with fineness modulus of 2.9. Table 3 shows the basic performance. Coarse aggregate is limestone with grain size of 5~25mm. There are three kinds labeled as G1, G2 and G3. The basic characteristics are displayed in Table 4.

Table 3. Characteristics of the sand.

Apparent density (kg/m^3)	Bulk density (kg/m^3)	Porosity (%)	Grading region
2670	1440	46.1	II

Table 4. Characteristics of course aggregate.

Number	Apparent density (kg/m^3)	Bulk density (kg/m^3)	Needle flake content(%)	Moisture content (%)	Cumulative percentage retained (%)					
					25	20	16	10	5	2.5
G1	2740	1380	7.4	0.80		11	40	86	100	
G2	2740	1420	4.2	0.63		5	26	76	100	
G3	2710	1380	6.1	0.70	2	16	37	87	99	100

2.2 Preparation of concrete

Through many previous studies on the influence of NC-H content, water-cement ratio, ratio between mortar and coarse aggregate and sand ratio on the concrete by our team, the mixing ratio of C60~C90 concrete was finally determined as shown in Table 5. P.O 42.5 cement was used for C60, C70 and C75 concrete, whereas, P.O 52.5 was used for C80 and C90 concrete.

2.3 Characterization

The slump and slump slow are conducted according to National Standard "Standard for test method of performance on ordinary fresh concrete" (GB50080). Compressive strength tests were carried out by an electro-hydraulic servo press (MTS YAW-4306, USA). The determination of compressive strength was performed after 1, 3, 7, 28, 60, 180 and 360d of hydration. Each resultant value of compressive strength was an average calculated from three determinations.

The sulphate resistance test of concrete refers to "Standard for test method of long-term performance and durability for ordinary concrete" (GB/T 50082).

The specimens were put in the fully automatic concrete sulfate dry-wet cycle tester (CABR-LSB). The solution was soaked in sodium sulfate solution with 5% concentration. After 150 cycles as required by the standard, the compressive strength of the specimen was tested.

In the freezing and thawing cycle test, the concrete rapid freeze-thaw tester machine (KDR-V) was used. Transverse dynamic modulus of elasticity was tested by dynamic elastic modulus tester (DT-W18).

The corrosion properties of reinforcement were studied by electrochemical accelerated corrosion test. The SAC-10 corrosion tester was used to conduct the experiment.

Table 5. Proportions of concrete mixture.

Concrete	Blending materials(kg/m^3)			sand ratio(%)	W/B
	Total	Cement	NC-H		
C60	520	390	NC-H60 130	38	0.28
C70	570	399	NC-H70 171	38	0.27
C75	580	377	NC-H70 203	38	0.27
C80	580	406	NC-H80 203	38	0.25
C90	580	348	NC-H90 232	38	0.25

3. Results and discussion

3.1 Working behavior of freshed concrete

The slump is the comprehensive performance of cohesion, fluidity and stability of concrete mixture. Generally, the technical performance of concrete mixture is controlled by the size of slump on the construction site. The experimental results are shown in Table 6. The initial slump of concrete is above 200mm, and the slump expansion is above 500mm. At the same time, the slump loss was slow, and the retention value of slump for 30min is around 200mm. This infers that the prepared concrete has a low loss of slump, no bleeding and segregation, which can well meet the needs of the construction technology of concrete on-site mixing, centralized mixing and tandem pouring. The concrete has good working performance.

Table 6. The workability of concrete.

	Slump (mm)		Slump flow (mm)
	0min	30min	
C60	225	200	520
C70	220	200	515
C75	210	195	510
C80	220	205	520
C90	210	190	505

3.2 Compressive strength

It can be seen form Figure 1 that the compressive strength of prepared C60~C90 concrete are all high in various ages. Among them, strength of 1d reaches 38% ~ 51% of the designed strength, and the strength of 3d reaches 76% ~ 86% of the designed strength. After 7 days, the strength reaches more than 100% of the designed strength. The strength of 28d reaches more than 112% of the designed strength. Besides, with the extension of the curing period, the strength at later stage continues to increase. When it reaches 1 year, the strength increased by over 8% compared with the strength at 28d. It is concluded that anticorrosive concrete prepared in the present study has good mechanical properties.

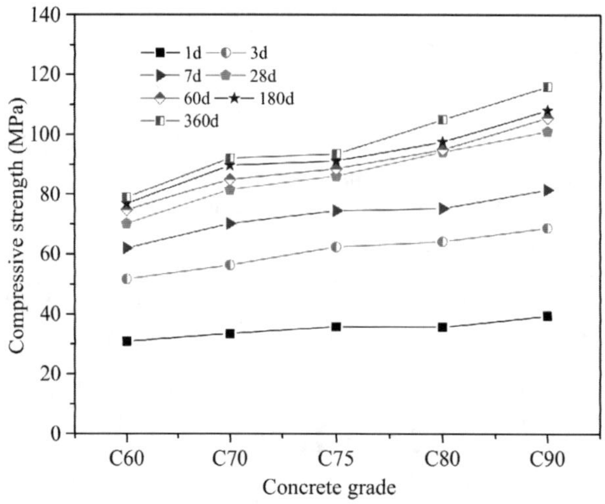

Figure 1. Compressive strength of anticorrosive concrete.

3.3 Dry shrinkage

Due to the characteristics of shaft construction, it is difficult to maintain the concrete. Therefore, the concrete shrinkage value measured after 3d in standard curing condition is quite different from actual situation. Here, the concrete is directly put into the curing room with constant temperature and humidity after 18h of concrete molded. Then the concrete shrinkage value is measured at different times which is shown in Figure 2. As can be seen, overall, dry shrinkage increases faster at the first 28 days and later it develops slower and tends to be stable. The dry shrinkage value of 1year is less than 5×10-4.

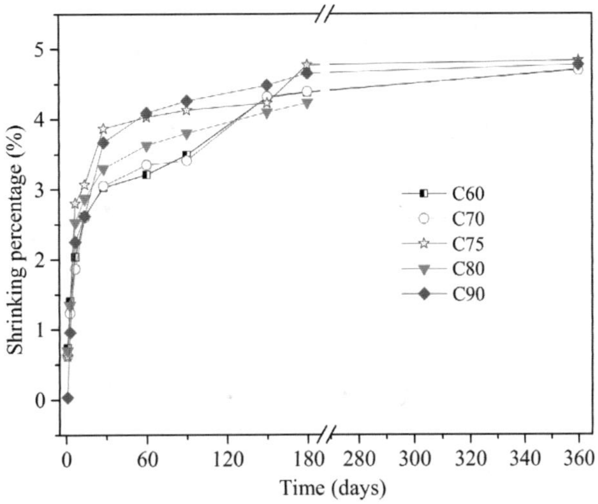

Figure 2. Shrinking percentage of prepared concrete.

3.4 Anti-permeability

It was well known that the transmission of corrosive ions in concrete seriously affects the durability. The diffusion coefficient of corrosive media in concrete can reflect the permeability of concrete. The Cl^- diffusion coefficient is investigated. It has been reported that the Cl^- diffusion coefficient of concrete with poor quality is in the order of $1000\times10^{-14}m^2/s$, whereas, in ordinary concrete the Cl^- diffusion coefficient is in the order of $100\times10^{-14}m^2/s$, and the Cl^- diffusion coefficient of concrete with high impermeability is in the order of $10\times10^{-14}m^2/s$ [16]. The relationship between Cl^- diffusion coefficient and concrete permeability is shown in Table 7.

Table 7. Relationship between Cl^- diffusion coefficient and concrete permeability [17]

Coefficient $10^{14}m^2/s$	>1000	500~1000	100~500	50~100	10~50	5~10	<5
concrete permeability	extremely high	high	medium	low	very low	extremely	ignored

In this experiment, Cl^- diffusion coefficient after 28 days was conducted. Figure 3 displays the results. It shows that the chloride diffusion coefficient of C60 concrete at 28d is 38.1×10-14m^2/s, and for C70~C80 concrete it is (10~20) ×10-14m^2/s, while that of C90 concrete at 28d is only 4.6×10-14m^2/s. In comparison with Table 7, it can be seen that the permeability of C60~C80 concrete at 28d is in a very low range, and that of C90 HPC at 28d is in an extremely low range. It firmly infers that the concrete has excellent anti-permeability. As the age of concrete rises, the compactness and permeability are both improved.

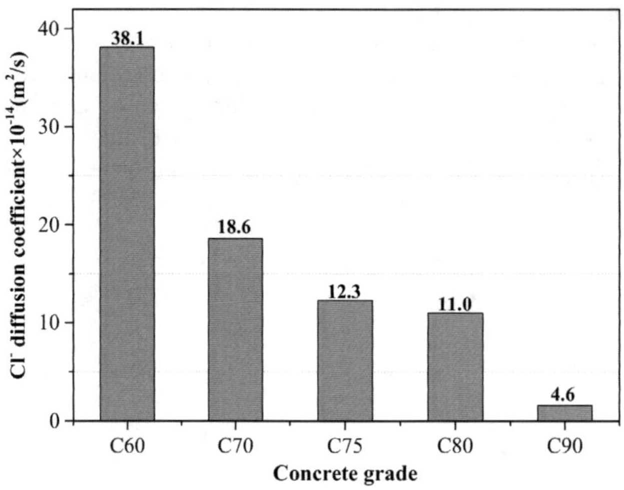

Figure 3. Chloride diffusion coefficient of prepared concrete.

3.5 Resistance to sulfate attack

Table 8. Results of sulphate attack test.

Concrete	Strength after 150 cycles	Standard curing strength at the same age	Corrosion resistance coefficient
C60	71.6	71.4	1.00
C70	81.2	80.3	1.01
C75	89.0	87.3	1.02
C80	97.8	94.2	1.04
C90	106.3	101.1	1.05

When concrete is in the environment containing SO^{4-}, which can react with $Ca(OH)_2$ of cement to generate $CaSO_4 \cdot 2H_2O$ and further ettringite occurs. This can make the solid phase volume in the hardened cement stone increase a lot, resulting in considerable crystallization pressure and leading to the cement stone cracking. The mechanical properties of concrete are reduced and destroyed finally. In general, the concrete first falls off at the edges and corners. As the SO^{4-} corrosion goes deeper, the concrete becomes brittle and loose. The results of C60~C90 concrete after 150 cycles is given in Table 8. It shows that the coefficient of resistance to sulphate attack gradually increases with the concrete grade increasing. All is not less than 1.00, and for C90, it reaches 1.05. So, it can be concluded that prepared anticorrosive concrete has good resistance to sulfate erosion.

3.6 Resistance to freezing and thawing cycle

Frost resistance, as an important indicator of concrete anticorrosion, can indirectly reflect the ability of concrete to resist environmental water intrusion and ice crystal pressure.

Fast freezing method is used to measure the frost resistance of C60~C75 concrete. The results are shown in Table 9. it can be found that the relative dynamic modulus of elasticity loss is less than 4% for C60~C75 concrete with 300 of freeze-thaw cycles. While the cycle times is 1000, the dynamic elastic modulus loss is less than 9% and weight loss is less than 3%.

In view of the small loss of C60~C75 concrete, slow freezing method is used to measure the C80 and C90 concrete, as is shown in Table 10. It can be discovered that, after 350 times of freeze-thaw cycles, the strength loss rate is only 5.3% for C80 concrete, and it is only 2.7% for C90 concrete. By comparing the results of fast freezing-thawing method and slow freezing method, it is found that slow freezing method is more severe. In general, prepared anticorrosive concrete in this work has good anti-

Table 9. Fast freezing-thawing cycle test of C60~C75.

Concrete	C60		C70		C75	
Times	Weight loss (%)	Relative dynamic modulus of elasticity loss (%)	Weight loss (%)	Relative dynamic modulus of elasticity loss (%)	Weight loss (%)	Relative dynamic modulus of elasticity loss (%)
200	0.39	2.94	0.49	3.32	0.15	3.03
300	0.39	3.30	0.69	3.90	0.35	3.50
400	0.49	4.22	0.74	5.6	0.35	3.62
500	0.98	5.06	1.14	5.78	0.50	4.31
600	1.47	5.14	1.33	5.85	0.60	5.06
700	1.57	6.31	1.73	6.36	0.80	5.36
800	1.97	6.99	1.93	6.71	1.00	5.63
900	2.26	7.65	2.27	7.19	1.30	6.14
1000	2.85	8.50	2.62	7.72	1.55	6.38

Table 10. Slow freezing-thawing cycle test of C80 and C90.

Concrete	Strength of 28d (MPa)	Freeze-thaw cycles of 350		
		Strength after freezing (MPa)	Standard curing strength (MPa)	Loss rate (%)
C80	94.2	99.9	105.5	-5.3
C90	101.1	109	112	-2.7

freeze and thawing cycling performance. This is mainly attributed to the NC-H admixture, which can improve the grading of pores and make the pore size smaller. The reduced porosity and structure dense are conducive to improve the ability to freeze.

3.7 Corrosion of reinforcement

There are two main reasons for the failure of steel passivation film in concrete: one is the carbonization of concrete; The other is the intrusion of corrosive ions, which diffuse onto the surface of the steel bar to blunt it. Oxygen and water must be involved in the corrosion of steel reinforcement, so the permeability of concrete is very critical to the corrosion of steel reinforcement.

The SCA-01 corrosion tester is used to evaluate whether concrete has corrosive influence on the steel bar, mainly by measuring the anode polarization curve of the steel bar in the hardened mortar and the current-time curve of the anode acceleration test under constant voltage. The anode current decreases with time, indicating that the anode process is inhibited and the steel bar will not be corroded. Figure 4 and Figure 5 gives the two types of curves.

It can be seen from Figure 4 that the potential gradually becomes stable with increase of current. It indicates that passivated film on the surface of steel bar is intact, and the steel bar in the concrete is in passivated state. The anode was then accelerated for 7 days. It is found that the anode current decreases with time (Figure 5). This infers that the anode process is inhibited to some degree, so that the steel bar will not corrode. Besides, it is noticeable that different sizes of current occur in all the concrete, this is due to the different inhabitation.

In a conclusion, prepared anticorrosive concrete of C60~C90 have good performance to inhibit corrosion of reinforcement. This may be attributed to the high impervious performance. Small Cl^- diffusion coefficient (Figure 3) can greatly delay the process of decarbonization by Cl^- erosion.

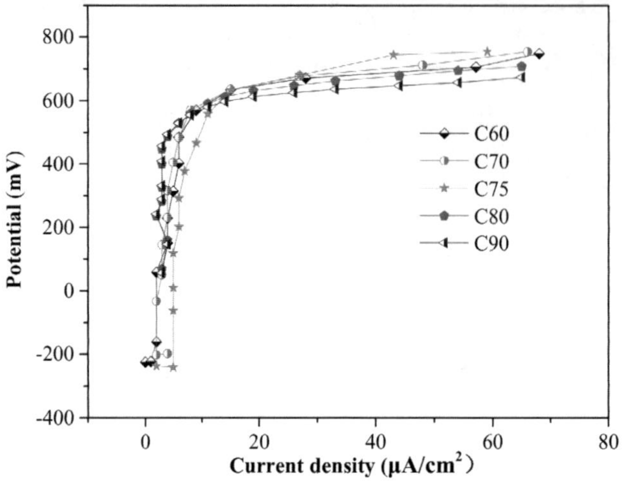

Figure 4. Anode polarization curve.

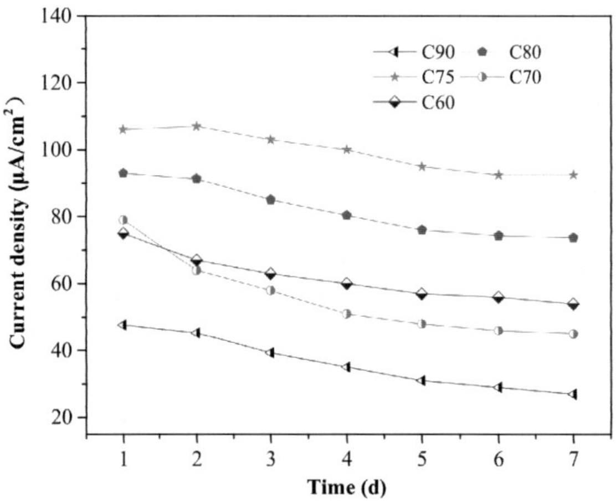

Figure 5. Current-time curve of the anode acceleration test.

4. Application

Anticorrosive concrete for shafts in thick and deep surface soil studied in this paper has been applied in some big coal mine projects. Here only two projects are introduced.

Longgu coal mine, as a national key project, has the highest comprehensive technical difficulty in Asia (Figure 6). The frozen section is designed with internal and external reinforced concrete shaft walls, with support thickness of 1000~2200mm. Concrete strength grade of C30~C70. The concrete in section -378m ~ 615m is C65~C70. The practical application proves anticorrosive concrete developed by our company has good construction performance, low slump loss, and fast developed concrete strength. At the same time, the strength of the later period continues to increase, completely meets the engineering requirements.

Guo Tun coal mine (Figure 7) has the freezing depth of 702m. C60~C75 anticorrosive concrete were adopted. No cracks appeared in the concrete shaft wall, and the quality of the project was very good after the inspection.

Figure 6. Longgu coal mine.

(a) (b)

Figure 7. Guo Tun coal mine.

5. Conclusions

From this study, it can be concluded that anticorrosive concrete for shafts in thick and deep surface soil are successfully prepared by use of conventional sand aggregate, ordinary Portland cement of grade 42.5 or 52.5, and especially NC-H compound admixture. The prepared C60~C90 concrete have high and good performance, as is shown in the following.

■ high working performance: The slump is above 200mm and have small loss, low viscosity and good construction performance.

■ high compressive strength: the strength at 1d reached 35%~60% of the designed strength; the strength at 3d reached 75%~95% of the designed strength; the strength at 7d reached more than 95% of the designed strength. The later strength continues to increase, after 1 year which is increased by more than 10% compared with that at 28d.

■ Low dry shrinkage: the dry shrinkage value of 1year is less than 5×10-4.

■ Good anticorrosion and durability: the permeability of chloride ion is quite low. Cl- diffusion coefficient: C60 is 38.1×10-14m2/s, while C90 is only 4.5×10-14m2/s; High resistance to sulfate corrosion with corrosion resistance coefficient of over 1.0; The high resistance to freezing and thawing circulation. After 1000 freeze-thaw cycles, the dynamic modulus loss of C60 was less than 8.5%, while that of C75 was less than 6.38%; good resistance to corrosion of reinforcement.

References

[1] Ye Y Q, Chu H Q, Jiang L H 2009 Durability and micro-mechanism of C70 high performance concrete in frozen shaft *Concrete* **8** 15

[2] Xu H D, Chu H Q, Wang M J, Xu N 2010 Development and application of high strength and high performance concrete to mine freezing shaft *Coal science and technology* **38** 38

[3] Peng S L, Rong C X, Cheng H, Wang X J, Li M J, Tang B, Li X M 2019 Mechanical properties of high-strength high-performance reinforced concrete shaft lining structures in deep freezing wells Advances in Civil engineering 2430652 1

[4] Yao Z S, Cheng Y, Xu H S, Liu S B 2006 Preparation and application of high strength and high performance concrete in freezing shaft wall of extra-thick surface soil *Coal industry* **8** 33

[5] Wang J, Chen J J, Ju Z Q 2013 Research on the mechanism and type of shaft lining reinforced concrete corrosion in salt disasters environment *Science technology and engineering* **13** 1690

[6] Niu X C, Yang R S, Jing L W 2004 Application prospect of high strength concrete in shaft lining *China mining* **13** 51

[7] Xu N 2007 Preparation and performance test of C100 high strength and high performance concrete Hohai university

[8] Shao C H, Yu H, Song L, Liu W 2007 Study on the ratio of high strength concrete to outer wall of deep surface frozen shaft *Mine construction technology* **28** 24

[9] Rong C X, Cheng Y, Yao Z S 2003 Repair and result of mine shaft lining in North ventilation shaft of Baodian mine *Coal science and technology* **31** 47

[10] Yang J J, Yao Z S, Liu Q L 2001 Design and construction of repairing and reinforcing auxiliary shaft of yokogawa mine *Mine construction technology* **22** 1

[11] Yao Z S, Cheng Y, Rong C X 2002 Optimization of shaft wall repair and reinforcement design of linhuan mine *Coal industry* **2** 4

[12] Yao Z S, Cheng Y, Yang Z H 2002 Analysis of mechanical mechanism of shaft wall failure in qujiang fengjing and repair and treatment *Coal science and technology* **30** 12

[13] Wu Z W, Lian H Z 1999 High performance concrete, China railway publishing house

[14] N M Ahmed, M G Mohamed, M R Mabrouk 2016 Study on the anticorrosive behavior of new core-shell pigments for protection of reinforced concrete steel in sulfate environment *Construction and building materials* **118** 226

[15] Theodosia Zafeiropoulou, Eleni Rakanta, George Batis 2013 Carbonation resistance and anticorrosive properties of organic coatings for concrete structures *Journal of surface engineered materials and advanced technology* **3** 67

[16] Guo L, Lu T W, Gu Q, Liu Z G 2007 Proerties and engineering application of grouting material prepared with high-performance grouting admixture *Insduatrial construction* **37** 93

[17] Liu Y B, Lu T W, Wang Q 2012 Performance and application in rigid waterproof engineering of concrete waterproofing agent *Concrete* **7** 86

IOP Conf. Series: Materials Science and Engineering 1028 (2021) 012008 doi:10.1088/1757-899X/1028/1/012008

Mechanical characteristics and comfort evaluation under pedestrian vertical excitation of a thin-walled cold-formed steel footbridge

Jing Xu, Xijun Chen, Jing Li* and Zeyu Li

School of civil engineering and transportation, south China university of technology, Guangzhou, Guangdong province, 510641, China

*Email: cvjingli@scut.edu.cn

Abstract. Thin-walled cold-formed steel (TWCFS) structures have a great application prospect and are widely adopted in low-rise residential constructions due to great mechanical and economical properties. Since TWCFS member has low flexural rigidity and is hard to weld, it is rarely used in bridge engineering. In Poyang Lake Wetland of Jiangxi Province, China, a TWCFS footbridge was built. In this paper, Serial tests including static loading test, natural frequency test and pedestrian comfort test were conducted. At the same time, finite element analysis (FEA) models were built with Midas/Civil software. The data shows that although natural frequency is far greater than the range of pedestrian walking frequency and will not lead to the resonance of bridge, the peak acceleration under crowd load reflects poor pedestrian comfort. The structure was modified for improving pedestrian comfort and the results after modification verify that the comfort can be ensured.

1. Introduction

Thin-walled cold-formed steel (TWCFS) structural members are normally fabricated from structural steel sheets by cold-rolling and brake-pressing methods [1]. TWCFS structures are becoming increasingly popular in construction industry, either as a complete or partial structure [2]. There are many advantages such as high ratio of strength-to-weight, convenient construction and less pollution [2-3]. Due to excellent mechanical and economical properties, TWCFS members have seen wide applications in low-rise residual structures [4-5]. However, due to the thin-walled characteristics, the flexural rigidity of TWCFS members is worse than that of general hot-rolled steel members, which makes the application of TWCFS in bridge design very rare.

Aluminum alloy is another material used in lightweight structures, but the elastic modulus of aluminum alloy material is only 1/3 of that of steel [6], and the flexural rigidity of members is significantly worse than that of steel members with sections of the same size. Nevertheless, there are still some aluminum alloy bridges in the world. The first aluminum alloy bridge was built in 1946 in New York, USA [7]. Canada built the first long-span highway aluminum alloy bridge around the world with a span of 88.4 m in 1949 [8]. In 2006, Hangzhou, China, built a number of aluminum alloy footbridges represented by the overpass at intersection of Qingchun Road and Yan'an Road [7]. Compared with aluminum alloy, the application of TWCFS in bridge field is almost zero.

In Poyang Lake Wetland of Jiangxi Province, China, a 13-span continuous footbridge made of TWCFS was built, designed as a bird-watching platform. The main structure of each span of this TWCFS footbridge weighs about 600 kg and is 11.2 m in length. Among them, the total weight of supporting members is about 186 kg, and the total weight of other members is 414 kg. Figure 1 shows some photos of this footbridge.

Figure 1. The TWCFS footbridge in poyang lake wetland.

Structures need to consider safety and serviceability at the same time. Safety refers to the ability to bear various possible loads and other actions, and serviceability refers to good performance under normal working conditions, which includes deflection, comfort and so on. For lightweight footbridges, even the safety requirements are met, people may feel uncomfortable due to excessive vibration when passing [9]. Xiebai footbridge, built in Hangzhou in 1993, was demolished in 2012 because of excessive vibration responses under pedestrian load [10]. The Millennium Bridge in London, which was built in 2001, was required to be stopped for use shortly after completion due to obvious vertical and lateral vibration. Some costly retrofitting had to be adopted to solve the vibration problems [11]. Therefore, the study of human-induced vibration is very necessary for the applicability of footbridges.

The problem of human-induced vibration of footbridges began to be of concerned in the 1970s. The national and regional codes mainly limit human-induced vibration by frequency adjustment method or combining it with dynamic response analysis method. The method of frequency adjustment is to avoid the possibility of resonance by adjusting the structure frequency not to fall into the range of pedestrian walking frequency. This method is simple and clear, adopted by the codes of most countries. The walking process of people consists of continuous steps, and the normal walking frequency is about 2 Hz [12]. In British bridge code BS5400 [13], the problem of human-induced vibration is not considered when the vertical natural frequency is greater than 5 Hz. The value reduces to 4 Hz in code OHBDC of Ontario, Canada [14]. Chinese code CJJ69-95 [15] requires that the natural frequency of the structure should be greater than 3 Hz.

In some codes, the dynamic response analysis method is adopted for footbridges which cannot meet the requirements of frequency adjustment method. This method evaluates the comfort of pedestrians directly by acceleration response, and the comfort limit value adopts peak acceleration or root mean square acceleration. However, there exists no necessity to analyze the acceleration response of the structure of which the natural frequency is beyond the range of pedestrian walking frequency. It is directly considered that avoiding resonance can meet the requirements of pedestrian comfort.

For the TWCFS footbridge located in Poyang Lake Wetland, some pedestrians reflected the lack of vertical comfort in use. Since each span of the structure is independent of each other, the static loading test and the natural frequency test were carried out based on a single-span structure, and the comfort response test was conducted at site. Finite element analysis (FEA) with Midas/civil software, which is a kind of FEA software and widely used in the field of bridge design in China was also used for studies during the process of tests.

2. Study of static and dynamic characteristics

Due to the lack of vertical comfort reflected by some pedestrians, tests for static and dynamic characteristics including the bearing capacity, deformation and natural frequency of the footbridge were conducted. According to Chinese code CJJ69-95, the ratio of deflection to span should be lower than 1/800; besides, the natural frequency shall be greater than 3 Hz. In the test, the independence of each span of the footbridge was considered, and the test was carried out based on a single-span structure.

2.1. Component composition and physical dimension

The TWCFS footbridge located in Poyang Lake Wetland is composed of 13 independent spans. During construction, each span structure shall be assembled firstly, and then transition trusses to realize the laying of additional wooden footpath boards connect the adjacent two spans. Each span is composed of one footpath truss and two handrail trusses on both sides. 12 connectors made of 3mm thick steel plate are used to enhance the connection strength of handrail truss and footpath truss. 8 among them are connected with the support rod at the same time. The other end of the support rod is fixed with the shallow foundation. The length of footpath truss of each span is 11.2 m, and the upper chord of each handrail truss with a length of 12.8 m is in contact with each other but not connected. Figure 2 and Table 1 show the component composition of a single-span and member section size.

2.2. Pre-study with FEA model

Before tests, analyses for static and dynamic characteristics were performed based on FEA model. In the model, only the TWCFS frame of the main structure was considered, and there was no additional wooden footpath board. Each component was rigidly connected at the position with bolts according to the design drawings. Referring to Chinese code CJJ69-95, the standard crowd load for bridges of which length is larger than 20 m is specified as:

$$W=\left[5-2\times\left(\frac{L-20}{80}\right)\right]\times(\frac{20\text{-}B}{20})(\text{kPa}) \tag{1}$$

Herein, W refers to the value of crowd load. L is the total length, while B is the bridge width. In formula (1), L should be taken as 100 m when the total length is greater than 100 m. According to the actual structure, L and B were taken as 100 m and 1.295 m respectively, and the calculated value of W was 2.8 kN/m^2, which was applied to double chords of footpath truss.

The structural static analysis showed that the maximum tensile stress was 34.6 MPa in the middle of upper chord, and the value of maximum compressive stress was 29.7 MPa in the middle of lower chord. Since the yield stress of Q235 steel is 235 MPa, all members have sufficient safety reserve. At the same time, the maximum deflection appeared at both ends of the structure with a value of 4.42 mm. Compared with the maximum limit value for the ratio of deflection to the span of 1/800 which is regulated in Chinese code CJJ69-95, that value of single-span model was 1/2534, which is completely satisfied.

The FEA model used subspace iteration method for modal analysis. The first-order frequency of the structure was calculated as 16.10Hz, which is much higher than the 3 Hz required by Chinese code CJJ69-95. At the same time, it completely avoids the high-order range of pedestrian step frequency, so resonance is almost impossible. Figure 3 shows the first four vibration modes.

2.3. Static and dynamic characteristic test

In static mechanical test, since the calculated crowd load was 2.8 kN/m^2, the total loaded weight was about 4 tons. Concrete blocks with an average weight of 75 kg were loaded in stages as shown in Table 2. Figure 4 shows the arrangement of strain gauges and displacement meters. Two displacement meters were arranged on both sides, and another was set in the middle.

As shown in Table 3, the stress and deflection under standard crowd load are compared with the value of former calculation with FEA model. In the test, the maximum tensile and compressive stresses were 37.70 MPa and 30.28 MPa, which appeared in the middle of upper chord and lower chord respectively. At the same time, the ratio of deflection to span was 1/2295 which is larger than 1/800 regulated in Chinese code CJJ69-95. The data obtained from test and Midas/Civil software were basically consistent, reflecting that the structure has a large reserve in bearing capacity and structural deformation.

For the actual multi-span structure, each span is connected to each other by transition truss. Thus, the free vibration of any single span is further limited, and it is safe to use the natural frequency of single-span structure to represent the lower limit natural frequency of the whole structure.

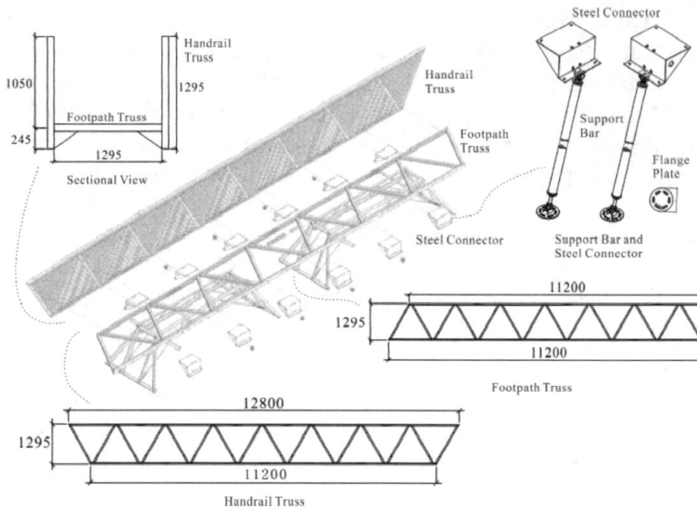

Figure 2. Component composition of one span footbridge.

Table 1. Section size of each member individually.

Component	Rod name	Section size(mm)	Edge size(mm)	Member length(mm)
Footpath truss	Chord	80×50×2	8	11200
	Web member	74×50×2	17	1475
Handrail truss	Upper chord	80×50×2	8	12800
	Lower Chord	80×50×2	8	11200
	Web member	74×50×2	17	1475
Support Bar	--	75×4.5	--	3000

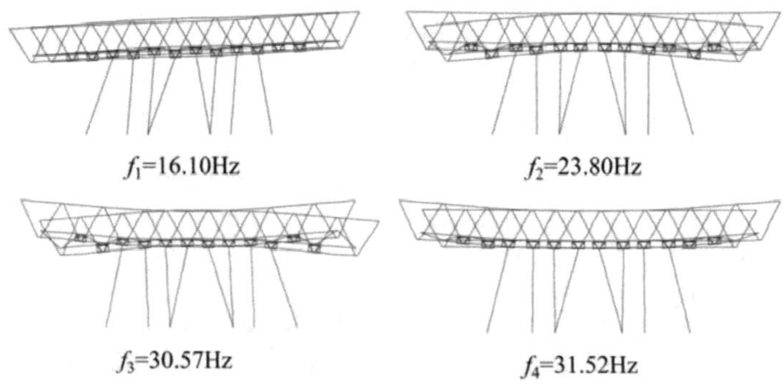

Figure 3. The first four vibration modes of the single-span footbridge.

Table 2. Graded loading system.

Load level	Concrete block mass (t)	Equivalent load (kN)	Loading ratio (%)	Rest duration (min)
Preloading	1	10	25%	5
First level loading	2	20	50%	10
Second level loading	2	20	100%	10

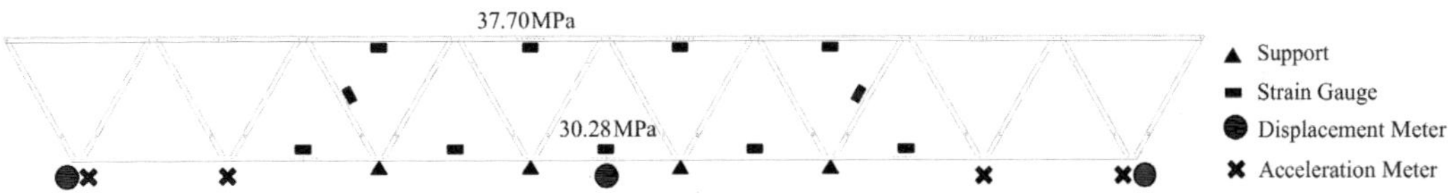

Figure 4. Layout of measuring points in static and dynamic tests.

Table 3. Results of static characteristic research.

Data source	Maximum tensile stress (MPa)	Maximum compressive stress (MPa)	Deflection (mm)	Deflection/Span
Midas/Civil	34.60	29.70	4.42	1/2534
Test	37.70	30.28	4.88	1/2295
Difference (%)	+8.96	+1.95	+10.41	-9.43

As shown in Figure 4, in the arrangement of acceleration sensors of singl-span structure, four acceleration sensors were symmetrically arranged at both ends of the structure. Five workers took a rope connected to one end of the bridge and jumped continuously to fully vibrate the single-span structure.

For analyzing the frequency spectrum of measurement points, Fast Fourier Transform (FFT) method was applied to the transformation of the spectrum from the time domain to frequency domain. Figure 5 shows the frequency spectrum of measurement point at the end of bridge with the most obvious vibration as a representative.

The measured spectra of other 3 measurement points are similar to Figure 5, and it is clear that the first two frequencies are 16.65 and 22.75 Hz respectively. In addition, these data are basically consistent with the first two frequencies of 16.10 and 23.80 Hz calculated by FEA model, which both illustrates that resonance will not occur for the TWCFS footbridge.

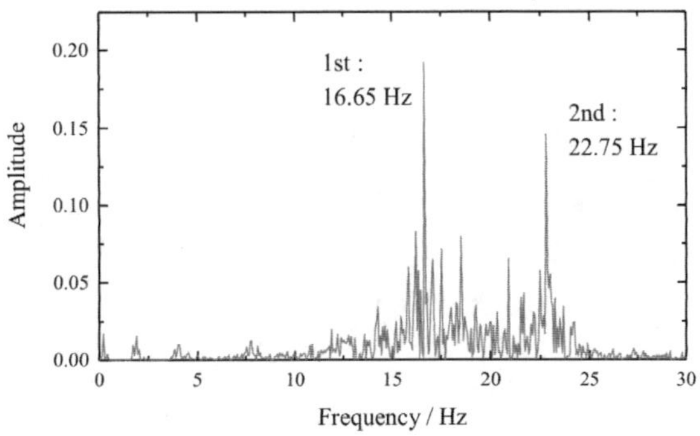

Figure 5. Frequency spectrum of the measurement point at the end of bridge.

3. Study of acceleration response of multi-span structure

The studies based on single-span structure show that the bearing capacity and natural frequency of the TWCFS footbridge meet the requirements of Chinese code, and there will be no structural damage and resonance. However, in view of some uncomfortable phenomena in actual use, acceleration response is used as direct criteria to evaluate the pedestrian comfort of the footbridge.

3.1. Test design and results

The single-span structure with a span of 11.2 m can be divided into 4.8 m long supported section and 3.2 m long side span on both sides. In the multi-span structure, the adjacent two spans only connect the two footpath trusses through a transition truss. It is considered that each span is independent of each other, and the side span position of each span is the most sensitive position of the structure. The test was carried out based on the multi-span structure at site, and two acceleration sensors were set at the position as shown in Figure 6 due to the restriction of test condition.

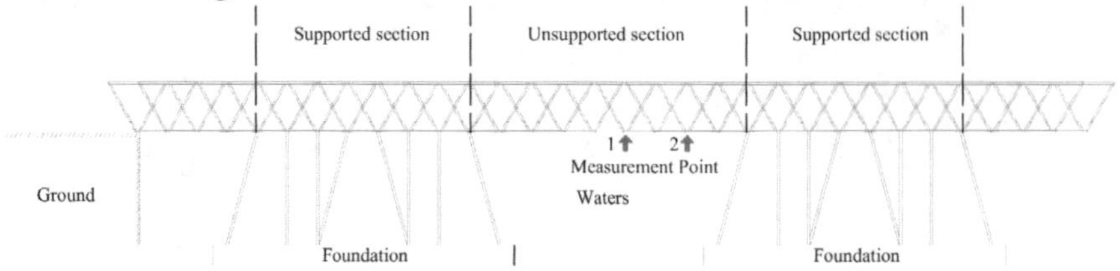

Figure 6. Schematic diagram of acceleration measurement test.

The footbridge is designed for scenic spot observation and photography, and is mainly loaded by pedestrians, and one walking cycle includes two steps. The landing of two feet on both sides corresponds to two peak walking forces, and the center of gravity is the lowest at this time. In the process of landing from the front foot to the back foot, the walking force reduces as the center of gravity is increased. With the increase of walking speed, the interval between the two peaks decreases until it overlaps in running state.

The actual walking process is random, and the walking frequency, peak value and phase vary from person to person. Random structural response cannot be directly superposed, and the maximum instantaneous step force of the total structure is less than the sum of peak values. According to the theory of random vibration, the dynamic response of N excitations with the same amplitude but the random phase is equal to $(N)^{1/2}$ times of the input response of a single excitation to a linear structural system. Therefore, there is a conversion relationship between the number of random walkers and the number of consistent walkers as shown in formula (2). When the crowd density is large (more than 1 person/m^2), pedestrians cannot walk according to their wishes and habits. The structural response should be magnified by 1.85 times to obtain the conversion relationship as shown in formula (3) when the crowd density is greater than 1 person/m^2 [16].

$$N_P = (N)^{1/2} \tag{2}$$

$$N_P = 1.85(N)^{1/2} \tag{3}$$

Among them, N_P is the number of pedestrians walking in consistent, while N is that of random pedestrians. The value of N can be calculated from $N = S \times L \times B$, in which S, L and B are the crowd density, the length and width of the bridge respectively.

In test, three walkers with similar weight and an average value of 60 kg walked across the bridge, and ensured that all pedestrians walked with the same phase. The measured walking speed is V=1.4 m/s on average. Combined with the relationship V= 0.71f_p proposed by Aikaterini [17], the walking frequency f_p is approximately 2 Hz, which is the same as that proposed by Bachmann et al. [12]. Figure 7 shows the acceleration time history response, and Table 4 is the peak data.

There are many different codes for pedestrian comfort indicators. EN03 [18] in Germany divides comfort level by peak acceleration as shown in Table 5, which is widely used.

The relatively large value between the absolute value of the maximum acceleration and the minimum acceleration is selected, and the pedestrian comfort is evaluated with reference to Table 5. Under the condition of single pedestrian walking, the peak acceleration is 0.34 m/s^2, which is at a very comfortable level. The peak accelerations of double and triple pedestrians walking are 0.69 and 0.87 m/s^2 respectively, which are in the medium comfortable range, but the peak acceleration of triple pedestrians walking is close to the lower limit of uncomfortable level 1.00 m/s^2. Thus, the comfort of the TWCFS footbridge under more pedestrian conditions remains to be studied.

3.2. Pedestrian load and FEA model

Based on the periodicity of pedestrian load, Fourier series pedestrian load model is widely used, which can be expressed by formula (4).

$$F(t) = G + G\sum_{i=1}^{\infty} \alpha_i sin(2i\pi f_p t - \varphi_i) \tag{4}$$

Table 4. Measured peak acceleration.

Number of pedestrians	Peak Acceleration at measuring point 1 (m/s^2)		Peak Acceleration at measuring point 2 (m/s^2)	
	Maximum value	Minimum value	Maximum value	Minimum value
1	0.33	-0.34	0.21	-0.24
2	0.67	-0.69	0.50	-0.48
3	0.87	-0.83	0.59	-0.61

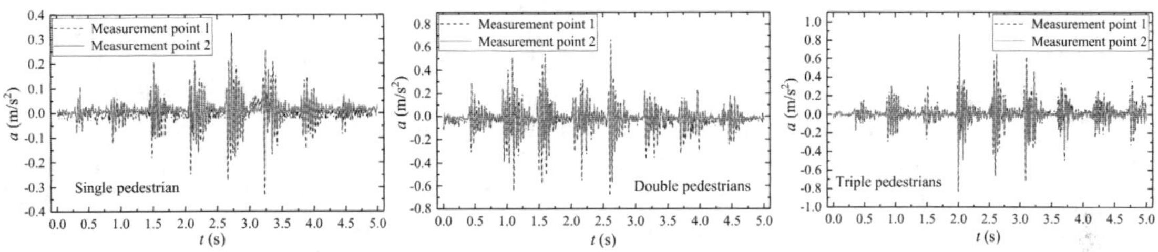

Figure 7. Measured acceleration time history.

Table 5. Comfort level specified in EN03 code [18].

Grade	Comfort level	Vertical acceleration limit (m/s^2)
1	Very comfortable	<0.50
2	Medium comfortable	0.50-1.00
3	Uncomfortable	1.00-2.50
4	Unbearable	>2.50

Among them, G is the weight of pedestrians, t is the time, and f_p is the walking frequency. α_i is the ith-order dynamic load factor, which is the ratio of harmonic amplitude value to pedestrian weight, and the ith-order harmonic phase angle is represented by φ_i.

Bachman et al. [12] proposed that the first-order dynamic load factor was 0.37 at 2 Hz walking frequency. Young [19] thought that the first-order dynamic load factor at 2 Hz walking frequency was 0.43. IABSE [20] stated that the first-order dynamic load coefficient was 0.4, while α_2 and α_3 were 0.1 at the walking frequency of 2 Hz. In addition, $\varphi_1 = 0$, $\varphi_2 = \varphi_3 = \Pi/2$.

According to the actual structure, a three-span model shown in Figure 8 was established on the basis of the single-span model. The number of pedestrians on the bridge deck was controlled to be 1-16 in sequence, and the pedestrians kept the same amplitude and phase to walk from left to right with space of 0.8 m between the front and rear pedestrians. The pedestrian weight G was taken as 700 N to simulate the acceleration response of the structure under pedestrians walking. Table 6 shows the calculation results, in which N_P is the number of pedestrians. The acceleration at the position of measurement point 1 in test was taken as the structural response, and a_{max} represented the peak value.

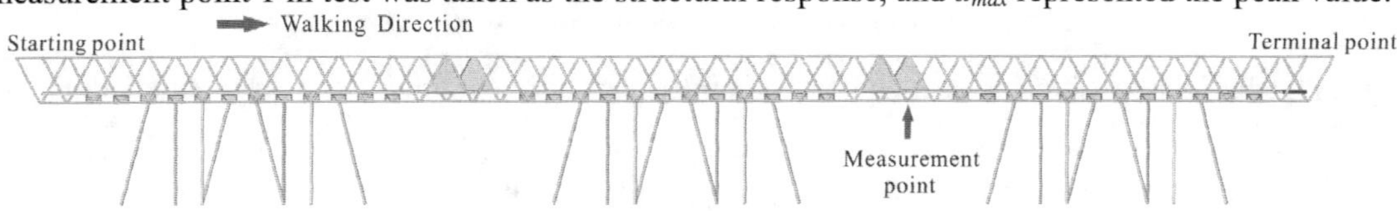

Figure 8. Three span FEA model.

Table 6. Maximum structural response of 1-16 pedestrians walking synchronously.

N_p	a_{max} (m/s^2)	N_p	a_{max} (m/s^2)	N_p	a_{max} (m/s^2)	N_p	a_{max} (m/s^2)
1	0.409	2	0.808	3	1.140	4	1.462
5	1.696	6	1.828	7	1.920	8	1.902
9	1.808	10	1.689	11	1.636	12	1.714
13	1.761	14	1.779	15	1.795	16	1.783

The results show that the maximum acceleration response value of the structure appeared when 7 pedestrians walked, and when the number of people was more than 9, the response value decreased gradually. Figure 9 shows the pedestrian position corresponding to the peak acceleration under various working conditions. When there were few pedestrians, the peak acceleration had a linear superposition relationship, which is similar to the measured acceleration response in test. It is reasonable that the peak acceleration calculated by the model is slightly larger than the measured results, considering that the absolute synchronous walking cannot be realized and the average weight of 60kg is lower than that in model. When the number of pedestrians increased to 7, the acceleration response at measurement point 1 continued to increase. All pedestrians were just full at the unsupported section when there were 11 pedestrians. When the number of pedestrians was more than 11, the peak acceleration response was stable between 1.7-1.8 m/s^2, reflecting that the pedestrian load in the supported section has no significant contribution to the response of the sensitive position of the side span of the structure.

The unsupported section is taken for independent analysis, which is composed of two side spans of 3.2 m long and 1.6 m long transition truss, with a total length of 8m. Under the condition corresponding to the most unfavorable load of the structure, all pedestrians walk in the unsupported section, therefore, L is taken as 8 m in the calculation of N in formulas (2) and (3). EN03 stipulates the largest crowd density is 1.5 person/m^2 [18], which is considered as a typical high-density crowd walking condition. According to formula (3), it is converted to 7.4 persons walking condition. In Figure 9, the interval between adjacent pedestrians is 0.8 m, which is more conservative than that of 7.4 people uniformly distributed along 8 m. Based on the calculation results of 7 and 8 pedestrians, the peak acceleration of 7 pedestrians walking at 0.8m interval is 1.920 m/s^2, corresponding to the uncomfortable interval in EN03 code. Under the density of 0.5 and 1.0 person/m^2, it is converted to 2.28 and 3.21 pedestrians walking uniformly and synchronously along the length 8 m according to formula (2). For safety, the acceleration response of 3 people and 4 people walking at 0.8m interval in Figure 9 is used to represent the random walking condition of 0.5 and 1.0 person/m^2. The corresponding maximum acceleration response is 1.140 and 1.462 m/s^2, which are slightly higher than the upper limit of the medium comfortable range and at an uncomfortable level.

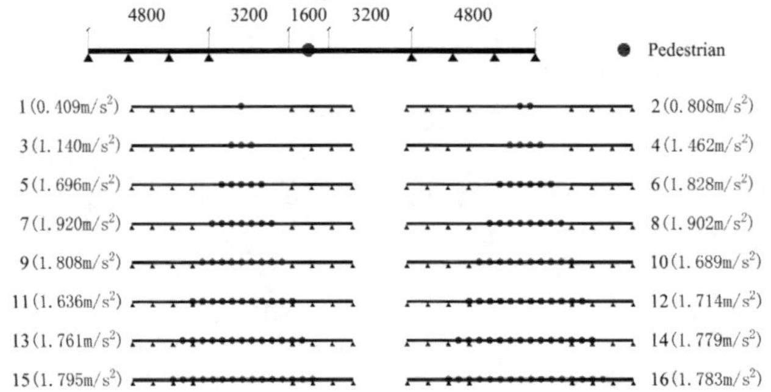

Figure 9. Corresponding pedestrian position under peak acceleration of each working condition.

According to the comprehensive measurement of test and FEA, the natural frequency of the footbridge is beyond the range of walking frequency. Although the possibility of resonance is avoided, it is not enough to completely evade the uncomfortable phenomenon of pedestrians. The design of independent supports for each span is adopted, and the supported section is not sensitive to human-induced vibration. However, the problem of insufficient vertical stiffness may exist in the section without support, resulting in the uncomfortable phenomenon of pedestrians.

4. Structural reinforcement

The control system represented by a tuned mass damper (TMD) is usually used when the pedestrian comfort is not satisfied for footbridges [21]. For the lightweight and simple TWCFS footbridge, TMD is not easy to arrange. Formula (5) is the forced vibration equation of damped multi-degree of freedom system.

$$[M]\{\ddot{y}\}+[C]\{\dot{y}\}+[K]\{y\}=\{P\} \tag{5}$$

Under the premise of constant external load $\{P\}$, since the mass matrix $[M]$ and damping matrix $[C]$ are not easy to change, the method of adding four support bars to the unsupported section as shown in Figure 10 was adopted for reinforcement to improve the vertical stiffness and reduce the pedestrian discomfort of structure.

IABSE load model was used to calculate the acceleration response of 1-16 pedestrians walking synchronously at 0.8 m spacing. The pedestrian weight G was taken as 700 N, and Table 7 shows the calculation results. N_P is the number of pedestrians walking, and the maximum acceleration response is represented by a_{max}.

The relationship between the peak acceleration of the structure and the number of pedestrians before and after the support reinforcement is similar. It reached the peak value when 8 pedestrians walked on the reinforced supported section and reached the minimum value when 12 pedestrians were covered between the two supported sections. After that, the peak acceleration slightly increased with the increase of the number of pedestrians in the supported section. According to the formulas of conversion from random walking people to the consistent walking crowd, the peak acceleration of 3 people walking synchronously is 0.610 m/s^2, which represents the response of crowd density 0.5 person/m^2, and it is in the medium comfortable range according to EN03 code. The peak acceleration response of crowd density 1.0 and 1.5 person/m^2 corresponds to the acceleration of 3 and 8 people walking synchronously, and they are also at the medium comfortable level.

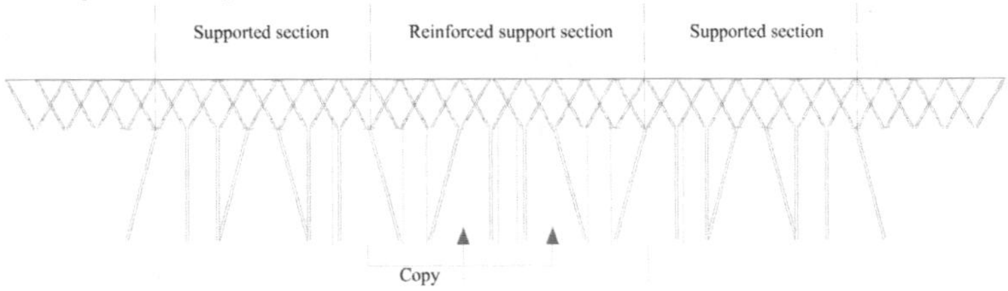

Figure 10. Arrangement of reinforced support bars.

Table 7. Maximum structural response of synchronous pedestrians after support reinforcement.

N_p	a_{max} (m/s^2)	N_p	a_{max} (m/s^2)	N_p	a_{max} (m/s^2)	N_p	a_{max} (m/s^2)
1	0.236	2	0.439	3	0.610	4	0.763
5	0.874	6	0.936	7	0.956	8	0.961
9	0.926	10	0.881	11	0.850	12	0.837
13	0.843	14	0.858	15	0.872	16	0.882

5. Conclusions

Based on the static and dynamic tests of the single-span TWCFS footbridge, it was considered that the bridge has no problems of bearing capacity and possibility of resonance. Then, the acceleration response of 1-3 pedestrians walking synchronously was measured, and the comfort level was evaluated to be at a very comfortable or medium comfortable level.

At the same time, the FEA model was built with Midas/Civil software. The calculation results in the single-span model were quite consistent with the experimental data, which provided a basis for further acceleration simulation of multi-span footbridge. After that, the acceleration response analysis of 1-16 people walking synchronously with equal spacing of 0.8 m was carried out. According to the conversion relationship of pedestrian number between synchronous walking state and random walking state, the comfort levels of 0.5, 1.0 and 1.5 person/m^2 crowd density reached uncomfortable level.

In order to improve the pedestrian comfort on the basis of the original structure, the method of support reinforcement in the unsupported section was adopted. The acceleration response of 1-16 pedestrians walking synchronously is similar to that before reinforcement. However, the acceleration response of the structure under the crowd density of 0-1.5 person/m^2 reached a medium comfortable level, which shows the feasibility of the reinforcement scheme.

After the structural reinforcement, the structure forms a continuous support, which not only compressed the space under the bridge, but also was not conducive to the development of this structure. Then, the supports were changed to a continuous simple support structure, and the vertical bending stiffness of the structure was modified by changing the height of the handrail truss. The comparison of three kinds of handrail truss height showed that the vertical stiffness improvement was effective for the improvement of pedestrian comfort.

References

[1] Cai Y C, Young B 2019 *Eng struct.* doi 10.1016/j.engstruct.2019.109331
[2] Zhang J F, Pang S Y, Gao J W, Deng E F, Wang H, Zhao J J 2020 *Thin wall struct.* doi 10.1016/j.tws.2019.106493
[3] Xu S H, Zhang Z X, Li R, Wang H 2019 *Constr. Build mater* **222** 1-14
[4] Dung S E, Li H R, Wen Q P 2015 *Thin. wall struct* **95** 161-169
[5] Feng R Q, Zhu B C, Xu P H, Qiu Y 2019 *Thin wall struct.* doi 10.1016/j.tws.2019.106238
[6] Mazzolani F M 1995 *Aluminium alloy structures*. london: chapman & hall
[7] Xu Y F 2007 Structural analysis of aluminum alloy pedestrian bridge. Hangzhou: Zhejiang university. (in Chinese)
[8] Xin Y B 2009 Structural design and mechanical properties of aluminum alloy girder. Changsha: Hunan university. (in Chinese)
[9] Setareh M, Gan S Q 2018 *J.* Perform constr fac. doi 10.1061/(ASCE)CF.1943-5509.0001213
[10] Chen J L 2007 Study on vibration comfort of pedestrian bridge in pedestrian excitation. Hangzhou: Zhejiang University. (in Chinese)
[11] Dallard P, Fitzpatrick T, Flint A J A, Bourva S Le, Low A, Ridsdill R M, Willford M 2001 *Struct eng.* **79** 17-33
[12] Bachmann H, Ammann H 1987 Vibrations in structures induced by man and machines. international association of bridge and structural engineering. zurich
[13] British standards institution (BSI) 1978 British standard specification for loads, steel, concrete and composite bridges, part 2. London
[14] Ontario Ministry of Transportation 1983 Ontario highway bridge design code. Ontario
[15] Ministry of Construction of the People's Republic of China 1996 Technical specifications of urban pedestrian overcrossing and underpass. Beijing. (in Chinese)
[16] Zhao P S 2011 Research on human-induced vibration characteristics and vibration control of light long-span structure. Changsha: central south university. (in Chinese)
[17] Aitaterini P, Ji T J 2005 *Struct. Eng* **83** 36-40
[18] Footbridge Guidelines EN03 2008 Design of footbridges guidelines human-induced vibrations of steel structures
[19] Young P 2001 Improved floor vibration prediction methodologies. *ARUP Vibration seminar* **10**
[20] Matsumoto Y, Nishioka T, Shiojiri H 1978 Dynamic design of footbridges. *Proceedings of the international association for bridge and structural engineering* **2** 1-15
[21] Caetano E, Cunha A, Moutinho C, Magalhaes F 2010 *Eng struct.* **32** 1082-1091

IOP Conf. Series: Materials Science and Engineering 1028 (2021) 012009 doi:10.1088/1757-899X/1028/1/012009

Research on the performance changes of the binder of the wear layer of open-graded asphalt mixture under different aging methods

Jianjun Tian[1,*], Xiao Shi[2], Hongpo Chen[2] and Shuzhen Ji[3]

[1] R&D engineer, Shanghai municipal engineering design institute (Group) Co., Ltd, Shanghai 200092, China;
[2] Senior engineer, Shanghai municipal engineering design institute (Group) Co., Ltd, Shanghai 200092, China;
[3] R&D engineer, Shanghai roadphalt asphalt technology co., ltd, Shanghai 200092, China

[*] Email:tianjianjun@smedi.com

Abstract. In order to study the effect of asphalt aging of open-graded friction course (OGFC) on the performance of high-viscosity styrene-butadiene-styrene(SBS) modified asphalt during the mixing of the asphalt mixtures and the whole life cycle, Thin Film Oven Test (TFOT), Rolling Thin Film Oven Test (RTFOT), and Pressurized Aging Vessel Accelerated Asphalt Aging Test (PAV) were used to conduct aging tests on the high-viscosity SBS modified asphalt. And then the aging of SBS modified asphalt is studied at two levels: microscopic and macroscopic. In the microscopic aspect, first use the fluorescence microscope technology to image the asphalt samples before and after aging. The above images were processed by MATLAB and Image-pro plus software, and the micro-structures of high viscosity SBS modified asphalt before and after aging were obtained. At the same time, use the DSR test to observe the changes in the rheological properties of the SBS modified asphalt before and after aging. On the macro level, the softening point, viscous toughness and pull strength are used to analyse the changes in the mechanical properties of the high-viscosity SBS modified asphalt before and after aging, and the aging conditions of the SBS modified asphalt under the micro and macro indexes are compared. The research shows that after short-term aging (simulation of asphalt aging during the mixing), RTFOT and TFOT the loss ratio of SBS in modified asphalt are at 78.5% and 68.3%, respectively, the degradation of SBS in modified asphalt is serious, and the SBS modified asphalt aging conditions under the macro and micro indicators are consistent. After long-term aging (simulating the aging of the asphalt during the whole life cycle), the loss ratio of SBS in modified asphalt is 83.4% (RTFOT + PAV) and 84.6% (TFOT + PAV). However, the macroscopic index of SBS modified asphalt shows obvious differences. The aging degree of SBS modified asphalt using RTFOT + PAV aging method is significantly higher than that of TFOT + PAV aging method, indicating that the performance of SBS modified was affected at the degradation of SBS during long-term aging, it is also significantly affected by the aging of the basic asphalt itself. Therefore, when mixing OGFC asphalt mixtures, the mixing time should be strictly controlled, and the mixing time should not be extended. OGFC asphalt mixtures should be maintained properly in time to keep their excellent road performance.

IOP Conf. Series: Materials Science and Engineering 1028 (2021) 012009 doi:10.1088/1757-899X/1028/1/012009

1. Introduction

In recent years, in order to improve the urban living environment and build high-standard ecological cities, new ideas for building a "sponge city" have been proposed successively throughout the country. Open-graded friction course (OGFC), as a new pavement with good road performance, can absorb vehicle driving noise, and at the same time can eliminate road area water in time to reduce water mist, which has been affected by cities in recent years. It is favored by builders. OGFC pavement has a higher porosity, which is generally required between 18 and 25%, so higher requirements are placed on the bonding performance and road performance of asphalt cements [1]. To ensure that the drainage asphalt mixture has better high temperature performance, low temperature performance, water damage resistance and durability, OGFC mixture needs to use high viscosity asphalt as the binder [2]. As a result, a new type of high-viscosity SBS modified asphalt specially used for OGFC pavement came into being.

For high-viscosity SBS modified asphalt, its aging has two effects: degradation of SBS polymer (decreased polymer molecular weight), and oxidation of asphalt (increased oxygen molecular weight and asphalt molecular weight) [3]. The final aging result is the combined effect of the two effects, and it is related to the type and dose of SBS [4]. This has led to a more complex aging process for SBS modified asphalt than matrix asphalt [5, 6]. In order to analyze the aging mechanism of SBS modified asphalt, dynamic shear rheometer (DSR), gel permeation chromatography (GPC), infrared spectroscopy (FTIR) and fluorescence microscopy are mainly used for research [7].

In this paper, in order to study the effect of asphalt aging and asphalt aging on the performance of high-viscosity SBS modified asphalt during the mixing of OGFC asphalt mixture, the asphalt samples processed by different aging methods were observed and analyzed with the help of fluorescence microscope, DSR and some conventional asphalt tests. Previously, it provided a reference for studying the aging characteristics of high-viscosity modified asphalt.

2. Experimental design

2.1. Test materials

Take the purchased high-viscosity SBS modified asphalt as raw material, and its performance index is shown in Table 1. The aged asphalts were obtained by the original asphalt processed according to the " Standard Test Methods of Bitumen and Bituminous Mixtures for Highway Engineering " (JTG E20), including the asphalt Film Oven Test (TFOT), the asphalt Rolling Thin Film Oven Test (RTFOT), and the Pressurized Aging Vessel Accelerated Asphalt Aging Test (PAV).

Table 1. SBS modified asphalt main technical indicators.

Items	Unit	Specification Requirements	Result
Penetration（25℃,100g,5s）	0.1mm	40-60	51.2
Ductility (15℃)	cm	≥20	35.8
Softening point	℃	≥60	93.2
Brookfield Viscosity (135℃)	Pa • s	≤3	2.31
Brookfield Viscosity (177℃)	Pa • s	-	0.75

2.2. Test methods

In order to analyze the performance changes of the high-viscosity modified asphalt under four different aging degrees, RTFOT, TFOT, RTFOT + PAV(RP), and TFOT + PAV(TP), from micro and macro dimensions, a variety of different test methods are used.

In the microscopic aspect, the standard method in the Dutch specification was used for sample preparation[8], using the OLYMPUS BX51 fluorescence microscope to collect fluorescence micrographs at 100 times, five samples were prepared for each asphalt, and each sample was taken 4-

6 photos. Then use MATLAB to perform depth of field expansion and grayscale processing on the collected images, and finally use image-pro plus processing to calculate a series of related micro parameters. In addition, DSR test of asphalt samples was conducted according to the JTG E20.

From a macro perspective, the asphalt samples were subjected to the asphalt rotational viscosity test (T0625-2011), the asphalt softening point test (T0607-2011) and asphalt fracture performance test (T0629-2011) in accordance with the JTG E20. The rotational viscosity test measured the Brookfield viscosity of asphalt and various aged asphalts at four temperatures of 135, 150, 165, and 180 ℃ respectively. The asphalt fracture performance test was conducted at -5 °.

3. Asphalt microstructure and performance changes

3.1. Fluorescence microscopy analysis

The microscopic image of the high viscosity SBS modified asphalt under the different conditions of RTFOT, TFOT, RTFOT + PAV, TFOT + PAV, as shown in Figure 1. The black is the asphalt phase and the white is the polymer phase.

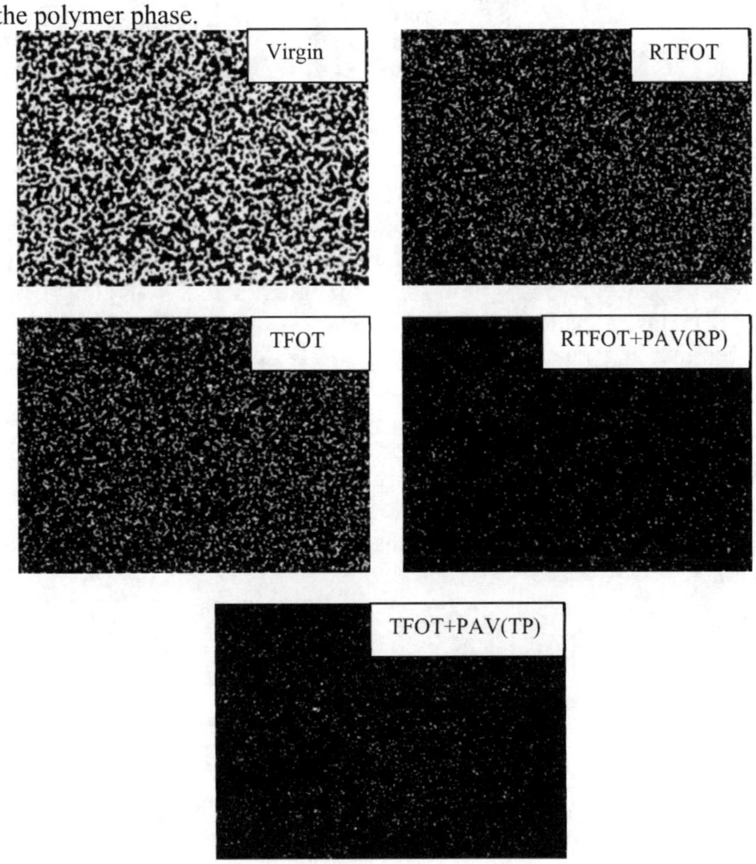

Figure 1. Microstructure of high-viscosity SBS modified asphalt before and after aging.

It can be seen from the microstructure diagram that in the original high-viscosity SBS modified asphalt, the SBS phase occupies most of the area, and the two phases are in a continuous state. There is also a small internal pore structure in the continuous SBS phase. After short-term aging (RTFOT and TFOT), it can be observed that the area occupied by the SBS phase has decreased significantly, and the previously continuous SBS phase has become a discontinuous island-like distribution. After the long-term aging (PAV), the area of the SBS phase continues to decrease, and the particle size of the dispersed SBS phase also decreases significantly compared to the short-term aging. This shows that whether short-term aging or long-term aging will cause the degradation of SBS polymer.

After processing and calculating the image with MATLAB and image-pro plus, a series of related micro parameters are obtained. Among these parameters, the SBS phase area ratio (Per Area), the average particle diameter (mean), and the average perimeter can intuitively reflect the characteristics of the SBS phase, but the SBS area ratio can most effectively characterize the degradation degree of the SBS phase. The area ratio of the SBS phase under different aging methods is shown in Figure 2.

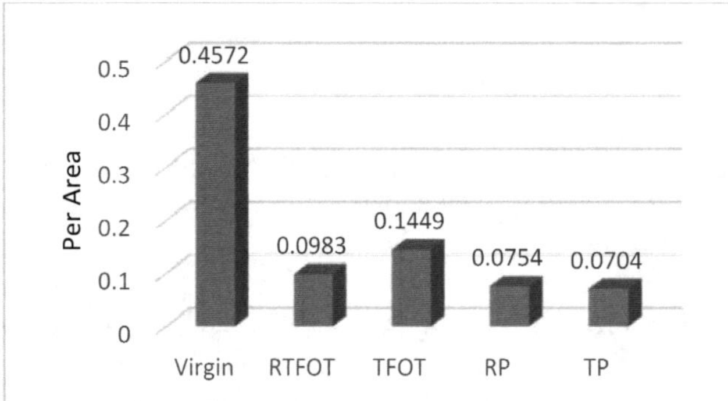

Figure 2. Changes of SBS phase area under different aging conditions.

It can be seen from Figure 2, both the long-term and short-term aging SBS phases have been degraded. The proportion of SBS loss after short-term aging is 78.5% (RTFOT) and 68.3% (TFOT); the proportion of SBS loss after long-term aging is 83.5%. (RTFOT + PAV) and 84.6% (TFOT + PAV), it can be seen that in general the degradation of SBS by long-term aging is greater than that of short-term aging, and the degradation of SBS by RTFOT is greater than TFOT in short-term aging.

3.2. Rheological analysis

Many studies have shown that there is a good correlation between the rheological properties of asphalt and the degree of asphalt aging [9, 10]. Therefore, this study conducted dynamic shear rheological (DSR) tests of high-viscosity SBS modified asphalt under different aging conditions, and the results are shown in Figure 3.

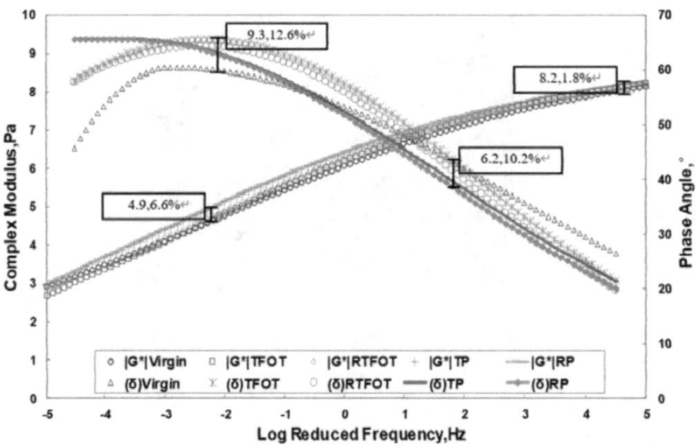

Figure 3. Master curve of SBS modified asphalt under different aging conditions.

It can be seen from the above figure, at the high position, the phase angle change amplitude is about 12.6%, and at the low position, this data is 10.2%. At the same time, in the high position, the change of the complex modulus is about 6.6%, and in the low position, this data is 1.8%. Different aging methods affect the phase angle and complex modulus of high viscosity modified asphalt, while the complex modulus of high viscosity modified asphalt is more affected by the aging method.

4. Macro index changes of aging asphalt

4.1. Kinematic viscosity

In order to study the viscosity change characteristics of high-viscosity modified asphalt after different aging degrees, the Brookfield viscometer was used. Using $\lg\lg\eta$ as the ordinate and $\lg(T + 273.13)$ as the abscissa, the corresponding points of the viscosity measured at different temperatures of the asphalt are plotted in the figure, and linear regression is performed on these points. The regression coefficient is the Viscosity Temperature Susceptibility (VTS) of the asphalt in this temperature range, that is, the viscosity temperature curve. The slope indicates the temperature sensitivity of the asphalt in the temperature range tested. Obviously, the smaller the VTS value is, the smaller the temperature sensitivity of the asphalt is. Related expression of the VTS are shown in formula (1).

$$\mathrm{VTS} = [\lg\lg(\eta_1 \times 1000) - \lg\lg(\eta_2 \times 1000)]/[\lg(T_1 + 273.13) - \lg(T_2 + 273.13)] \quad (1)$$

where: VTI - the viscosity temperature index by fitting.

η- The viscosity measured by the test.

T- The test temperature.

The viscosity-temperature curves of five different asphalts are shown in Figure 4, and the corresponding regression equations are shown in Table 2.

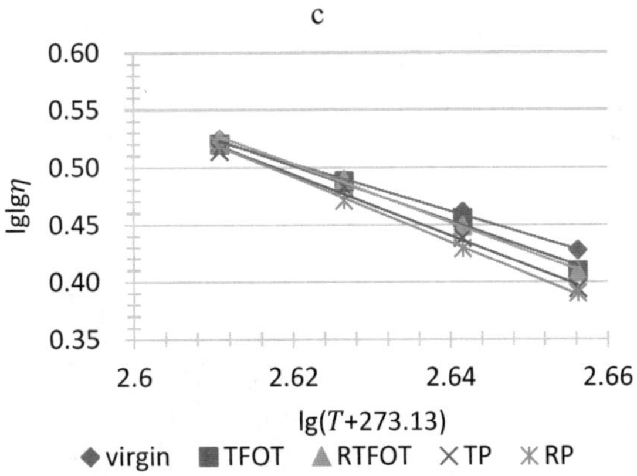

Figure 4. Viscosity-temperature curve of asphalt before and after aging.

Table 2. Regression equation of viscosity-temperature curve of asphalt before and after aging.

Asphalt	Viscosity-temperature relations	R^2
virgin	y=-2.1317x+5.8403	0.9951
TFOT	y=-2.4103x+6.6795	0.9898
RTFOT	y=-2.5623x+7.1653	0.9971
TP	y=-2.2346x+7.3462	0.9893
RP	y=-2.2456x+7.2435	0.9959

It can be seen from the viscosity-temperature curve of Figure 4, the temperature sensitivity of high-viscosity SBS modified asphalt gradually increases with the aging degree increasing, and the temperature sensitivity of modified asphalt after RTFOT + PAV aging is slightly greater than TFOT + PAV aging modified asphalt.

4.2. *Softing point*

For high-viscosity modified asphalt, the softening point is an indicator that is often used as a reference. The measured softening points of the high-viscosity asphalt under the original conditions and different aging degrees are shown in Figure 5.

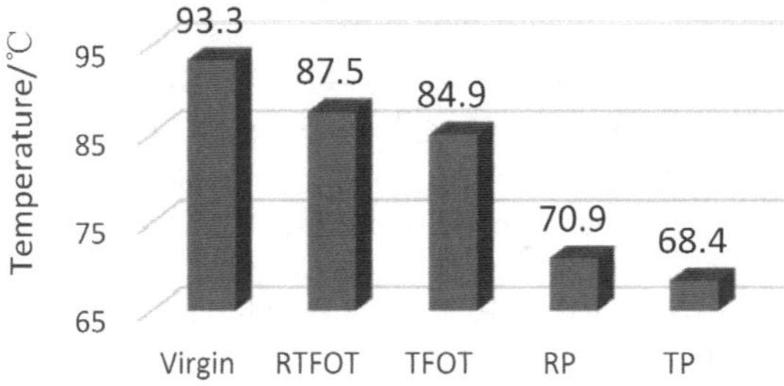

Figure 5. Softening point of asphalt before and after aging.

It can be seen from the above figure, the softening point of high-viscosity SBS modified asphalt has decreased after aging, whether it is short-term aging or long-term aging. Compared with short-term aging, the softening point of asphalt after long-term aging has decreased more. The decline is 3 times of short-term aging. For the short-term, the decrease in softening point of asphalt after RTFOT aging is slightly larger than that after TFOT aging.

4.3. *Pull strength*

In this paper, the tensile strength of the asphalt is determined by the direct tensile test as the pull strength of the asphalt. The results of the drawing strength are shown in Figure 6:

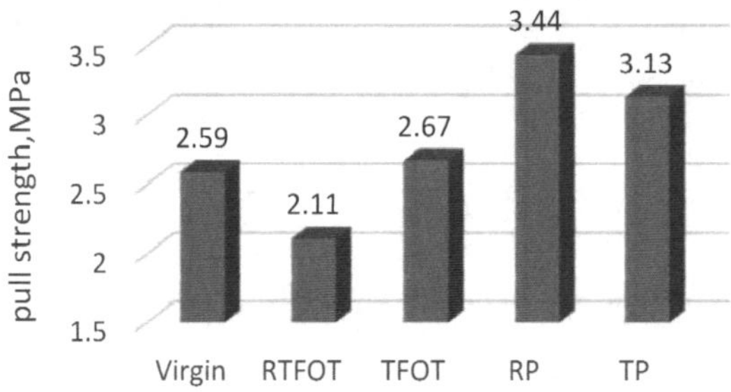

Figure 6. Pull strength of asphalt before and after aging.

It can be seen from Figure 6 that compared with the original bitumen drawing strength, the average drawing strength decreases after TFOT, but the average drawing strength increases slightly after RTFOT. And the average drawing strength after long-term aging (PAV) compared with the original asphalt, it has increased by 32.8%(RP) and 20.8%（TP）, respectively. The increase in pull strength after RTFOT + PAV is much larger than that of TFOT + PAV. This is because the modulus of asphalt will continue to increase after aging, so when the pull test is performed, the pull strength will continue to increase with aging.

5. Conclusions

By changing the asphalt aging method to prepare test material, based on the existing test equipment and methods, the influence of aging methods on SBS modified asphalt aging is explored. The main conclusions are as follows:

(1) Aging has significantly changed the microstructure of high-viscosity SBS modified asphalt, the area occupied by the SBS phase has decreased significantly, and the previous continuous SBS phase

has become a discontinuous island-like distribution. Long-term aging degrades SBS more than short-term aging. In short-term aging, the degradation of SBS is more in RTFOT than in TFOT.

(2) Observed by the fluorescence microscope image, after long-term aging (simulating the aging of the asphalt in the full life cycle), the loss ratio of SBS in modified asphalt is 83.4% (RTFOT + PAV) and 84.6% (TFOT + PAV). However, the macro indicators show significant differences. The reason for the small difference in microscopic performance between the two different aging methods may be because the viscosity of the high-viscosity modified asphalt used this time is not large enough to reflect the difference between the two aging methods in the high-viscosity asphalt high-viscosity state.

(3) Different aging methods all have an impact on the phase angle and complex modulus of high viscosity modified asphalt, and the complex modulus of high viscosity modified asphalt is more affected by the aging method.

(4) Analyzed from the macro index, with the intensification of aging, the temperature sensitivity of high-viscosity SBS modified asphalt is gradually increasing. After aging, the softening point will decrease. Compared with short-term aging, the softening point of asphalt decreases after long-term aging are more obvious. The aging degree of RTFOT is greater than that of TFOT. RTFOT aging method has a higher degree of aging for asphalt than TFOT, and this difference still exists after long-term aging.

(5) When mixing OGFC asphalt mixture, the mixing time should be strictly controlled, and the mixing time is not easy to extend. OGFC asphalt mixture should be cured in time to maintain its excellent road performance.

References

[1] Cheng Hua 2011 research on the application of OGFC for urban road drainage asphalt mixture Xi an: Doctoral thesis of changan university
[2] Zhu Man 2015 research on performance evaluation method and application of high viscosity modified asphalt guang zhou: Master thesis of south china university of technology
[3] Lu X, Isacsson U. 1998 chemical and rheological evaluation of ageing properties of SBS polymer modified bitumens *Fuel* **77** 961
[4] Chen Huaxin, Wang Bingang 2007 dynamic mechanical properties of SBS modified asphalt *Journal of south China university of technology: natural science edition* **35** 7
[5] Han Ling 2019 the structure and performance of Suizhong SBS modified asphalt *Petroleum Asphalt* **33** 4
[6] Huang Weidong, Sun Lijun 2002 mixing principle and process of SBS modified asphalt *Journal of tongji university: natural science edition* **30** 189
[7] Kang Aihong, Chen Juan, Sun Lijun 2012 performance and microstructure analysis of SBS modified asphalt *Journal of PLA university of science and technology :natural science edition* **6** 12
[8] Wegan V 1996 Sampling for microscopy-test procedure Danish road institute
[9] Wang Lan, Chen Gang, Xing Yongming, et al 2015 the effect of aging on the rheological properties of rubber powder and SBS modified asphalt *Journal of building materials* **3** 16
[10] Ouyang Jun, Sun Daquan, Zhang Yi 2011 research on the aging dynamic performance of SBS modified asphalt based on rheological properties *Journal of chongqing jiaotong university :natural science edition* **3** 30

Foaming effect influence of PBT / ABS resin as a reducing agent in foamed asphalt

Ming Huang[1,2,5,*], Hongpo Chen[3], Xiao Shi[4] and Xuejun Wen[5]

[1] Senior engineer(R&D), PhD, postdoctoral, greenland infrastructure group co, ltd, Shanghai 200010, China;
[2] Senior engineer(R&D), PhD, postdoctoral, GIG technology research institute, Shanghai 200010, China;
[3] Senior engineer, Shanghai municipal engineering design institute (group) co, ltd, Shanghai 200092, China;
[4] Senior engineer, Shanghai municipal engineering design institute (group) co, ltd, Shanghai 200092, China;
[5] Vice-chief engineer, Shanghai municipal engineering design institute (group) co, ltd, Shanghai 200092, China

*Email: huangming@tongji.edu.cn

Abstract. Foamed asphalt provides good environmental protection characteristics and economy. The utilization rate of old materials in foam asphalt recycling technology is higher than that of emulsified asphalt. Due to the lack of performance, the application scope of foam asphalt recycling technology is limited. In order to maintain high utilization rate of recycled asphalt concrete and improve the performance of foamed asphalt mixture, it will be the focus of research. In recent studies, adding a reducing agent into the regeneration process of foamed asphalt is a new technology. According to the test results of reduction performance of asphalt SARA four components, the PBT/ABS resin (Acrylonitrile butadiene styrene modified by Polybutylene Terephthalate) has excellent asphalt reduction effect, and PBT/ABS resin has good compatibility with emulsified asphalt after emulsification. Therefore, based on the existing foaming theory, the influence of asphalt temperature and water consumption on asphalt foaming with PBT / ABS resin reducing agent was designed. The influence of asphalt temperature and water consumption on foamed asphalt modified by PBT / ABS resin reducing agent was discussed. The results show that the PBT/ABS resin has larger foam volume than ordinary foam asphalt, but the average diameter of foam is slightly smaller. Similar to ordinary foamed asphalt, expansion rate and half-life are opposite to that of water consumption. Water consumption and asphalt temperature have significant effects on foaming effect. At the same temperature, with the increase of water consumption, the expansion rate increases and the half-life decreases. Under the same water consumption, the higher the foaming temperature, the greater the expansion rate. However, when the temperature reaches 160 °C, the expansion rate of asphalt does not vary with the temperature. With the increase of temperature, asphalt with different temperature has different optimal foaming water consumption. Finally, the optimum foaming conditions of PBT/ABS resin foam are given as follows: the best foaming temperature is 140 C, the water consumption is 3.8%, the ultimate expansion rate is 21.5%, and the half-life is 22.5s.

1. Introduction

At present, the development of road engineering has generally transited from the early construction stage to the current construction and maintenance stage. Therefore, from the perspective of economic and environmental benefits, in order to protect the environment and avoid the waste of resources, the recycling techniques of reclaimed asphalt pavement materials are gradually increasing[1]. Among them, cold recycling of foamed asphalt is the most economical method, and therefore widely concerned by researchers for its high RAP content ratio and rapid construction speed[2,3].

This research is based on the test section of a field construction. A lot of pre-research indicated that, the current aging degree of the binder in RAP, cannot be applied to the upper layer by ordinary foam regeneration technology. In order to use all recycled materials as much as possible, increasing the recycling rate becomes the path that must be passed [4,5]. PBT / ABS resin has become a potential good choice and entered the field of vision of researchers in this study.

PBT / ABS resin foam is made of PBT / ABS resin reductant, asphalt, water and compressed air. It is the same as ordinary foam asphalt. Asphalt is coated with steam bubbles, which causes asphalt to expand and increase the belly area of asphalt. During the process of asphalt foaming, the volume expands, and the asphalt has undergone obvious physical deformation. Aiming at the role of raw asphalt, the PBT / ABS resin reductant has increased the penetration rate and increased the content of flexible phenol [6-8]. Because there are few earlier stage studies on the foaming effect of foam asphalt added with special reducing agent, this study will mainly explore the temperature and utilization of content of water based on the existing foaming theory, at the same time, this study will provide some reference for the foam asphalt technology incorporating the reducing agent. The influence of water content on the foaming of asphalt with PBT / ABS resin reductant was emphasized, as well as the law of the change of expansion rate and half-life. This study selects the foaming index of PBT / ABS resin expansion foam and the half-life varying curves to comprehensively evaluate the foaming effect of asphalt, to determine the best foaming conditions.

2. Materials and experimental

2.1. Materials

PBT(Poly Butylene Terephthalate) was used to modify the conventional ABS(acrylonitrile butadiene styrene) to obtain PBT / ABS resin with good toughness. More importantly, its gum component can supplement the aged asphalt. The appearance of PBT / ABS resin is shown in Figure 1, and the appearance after shearing, dispersing and emulsifying is shown in Figure 2. The production method is not detailed, for it involves the unpublished patent and technical protection.

Figure 1. Appearance of PBT / ABS resins.

Figure 2. Emulsified resin reducing agent.

The principle of asphalt component test comes from liquid-solid adsorption chromatography. Adsorption is the accumulation of liquid molecules on the solid surface due to the attraction of unbalanced force. According to the different adsorption capacity of substances on the adsorbent, the separation effectives achieve. Generally, the substances with high polarity are easy to be adsorbed by the adsorbent, and the substances with weak polarity are not easy to be adsorbed by the adsorbent, so the components of asphalt can be separated. According to the China guideline JTGE-20-2011" Standard Test Methods of Asphalt and Bituminous mixture for Highway Engineering" [9]. The asphalt was separated by n-heptane, toluene and ethanol.

Three groups of experiments were conducted and the mean value was taken. The four component test results(SARA S(saturate), A(aromatic), R(resin), A(asphaltene)) are shown in Table 1.

Table 1. Results of SARA test.

Item	Fresh asphalt	Aged asphalt	Reduced asphalt	Coefficient of Variation 1 (%)	Coefficient of Variation 2 (%)
Asphaltene(%)	14.6	44.5	18.3	204.7945	25.34247
Resin(%)	28.8	23.3	26.5	19.09722	7.986111
Saturate(%)	30.6	18.8	31.2	38.56209	1.960784
Aromatic(%)	26	13.4	24	48.46154	7.692308

Where, Coefficient of Variation 1 refers to the difference of components between aged asphalt and fresh asphalt, and Coefficient of Variation 2 (%) refers to the condition after reduction.

Calculation method of Coefficient of Variation, see formula 1:

$$CV = |(A - A_0)\ /\ A_0| \times 100\% \tag{1}$$

2.2. Test method

Previous studies have shown that foamed asphalt has adopted the established index system, that is, Expansion ratio (ERm) and half-life (Half-life, 1/2) [10]. The ratio of the maximum volume to the original volume in the process of asphalt foaming is defined as the expansion rate. As the foamed asphalt is prepared as the process of the edge foaming edge decay, the measured expansion rate is slightly smaller than the true value. Under certain conditions, the larger the expansion area of the foamed asphalt, the better the dispersion of aggregates in asphalt. The half-life refers to the time it takes for the volume of the foamed asphalt to decrease to half of the maximum volume (with an accuracy of 0.1s). The longer the half-life, the slower the foam decay rate, that is, the half-life characterizes the relative stability of the foamed asphalt, so there is more time for mixing with mineral materials, effectively improving the performance of the mixture[11]. It will comprehensively consider the expansion rate and half-life to analyze the test results of PBT / ABS resin reduced asphalt foaming.

The expansion rate and half-life of the foam show the opposite trend with the change of water consumption. At the same time, the expansion rate and half-life have the lowest numerical limit. The best foaming condition should be determined according to the lowest numerical limit of the two. The middle value in the range should be selected as the best foaming condition at this temperature [12,13].

In addition, there is a fixed relationship between expansion rate and half-life. Foamed asphalt foaming is the process of edge foaming decay. Before measuring the expansion rate, the asphalt starts to burst [14], so we know that the maximum expansion rate is not the true expansion rate. Therefore, Jenkins uses differential method to get the general solution of ER_a formula 2.

$$\mathrm{ER}_a = \mathrm{ER}_m t_s(\ln 2)/\tau_{\frac{1}{2}} / \left[1 - \exp\left(-\frac{\ln 2}{\tau_{\frac{1}{2}}}\right) t_s\right] \tag{2}$$

Where: ER_m- expansion rate measured by test;

$\tau_{1/2}$—The half-life measured by the test;

t_s—The asphalt spraying time is 5s The amount of foaming test is generally 500g, and the spraying rate is about 100g / s, then 5s is taken.

In view of this, by studying the decay curves of foamed asphalt, the foaming index (Foaming index, FI) shown in formula 3 is used to evaluate the foaming properties of asphalt. Therefore, the expansion rate, half-life and foaming index are used to evaluate the foaming characteristics of asphalt.

$$FI = -\tau_{1/2}[4 - ER_m - 4\ln(4/ER_m)]/\ln 2 + (1 + c)ER_m t_s/2 \quad (3)$$

Where: - asphalt spraying time 5s;

C -- ratio of expansion value test value to true value.

Figure 3 shows the indoor foam asphalt testing machine chosen for this study. It is characterized by precise control and can adjust water volume, temperature and pressure.

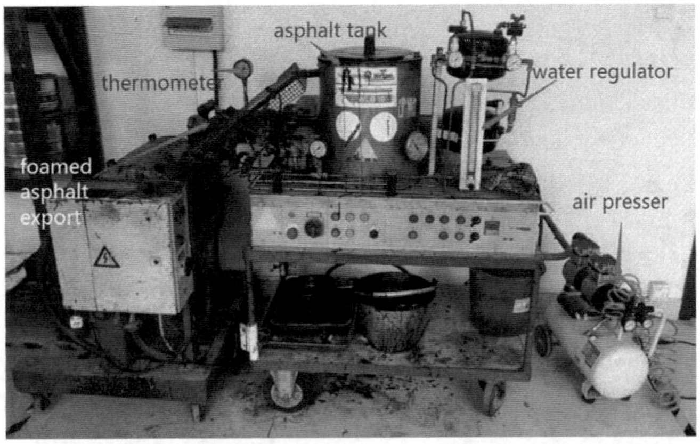

Figure 3. Witgens foaming equipment for test.

The addition of reducing agent affects the foaming degree and half-life. For decreasing uncertainty during the physical reactions, the emulsification of PBT / ABS resin is required to be very uniform. In addition, the steps of adding PBT / ABS resin and foam spray reducer should be early. In this test, a spray port was added. While fresh asphalt was added, the emulsified PBT / ABS resin was added into it. The amount of PBT / ABS resin reductant is 15wt% of total asphalt (fresh asphalt and recyclable effective asphalt). The foaming states of PBT / ABS resin foam are as Figure 4 and 5:

Figure 4. Appearance of common foamed asphalt(virgin asphalt).

Figure 5. Appearance of PBT / ABS resin reduced asphalt.

It can be seen from the appearance that the average diameter of foam is smaller than that of ordinary asphalt, but the amount of foaming is larger. Its binding effect needs to be further verified.

It is noteworthy that before the foaming test, the standard asphalt dosage is 500g, and the scale is erected on the side of the barrel. During the test, observe and record the maximum volume of asphalt expansion and the time when the volume is reduced to half of the maximum volume, namely the expansion rate and half-life respectively. Due to the greater influence of human factors in the experiment, each group conducted three parallel experiments. During data processing, error analysis is performed, and the test results with larger errors are discarded and the average value is taken.

2.3. *Experimental design*

Asphalt foaming is affected by water pressure, air pressure, asphalt type, asphalt temperature and water consumption [15]. In this test, the water pressure is set to 5bar, and the air pressure is set to 4bar. The rest will be studied as variables. After PBT / ABS resin reductant is added, the asphalt is not suitable for foaming below 130 °C, and above 170 °C, it is easy to cause reductant failure and further aging of asphalt. Therefore, in this test, the optional values of temperature as a variable are 130 °C, 140 °C, 150 °C, 160 °C and 170 °C. Based on the previous experimental experience and literature research[16], the available variables of water consumption are determined to be 2%, 2.5%, 3%, 3.5% and 4%. The orthogonal experimental design is shown in Table 2:

Table 2. Orthogonal experimental design

Influencing factors (variable indicators)	Temperature / ℃	130	140	150	160	170
	Water consumption	2%	2.50%	3%	3.50%	4%

In order to control error, all the above data have been tested in parallel for three times. In case of abnormal data, it is necessary to remove and conduct supplementary test. All the above test values are the average of the parallel test data obtained after removing the abnormal data and supplementary test. In the following sections, the data will be analyzed in detail through the trend chart.

3. Test results and analysis

3.1. *Effect of asphalt temperature on foaming effect*

The curves are drawn and studied for that the expansion rate and half-life will reach a value as large as possible[17]. According to the test plan in Table 2, the influence of asphalt temperature on the expansion rate and half-life are respectively plotted, as shown in Figure 6 and 7:

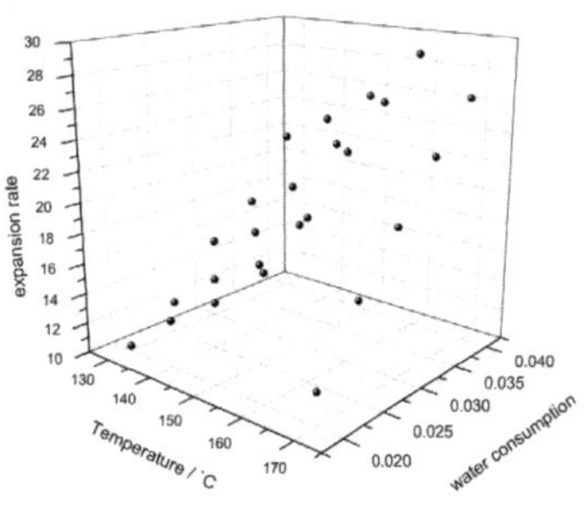

Figure 6. Expansion rate of foamed asphalt.

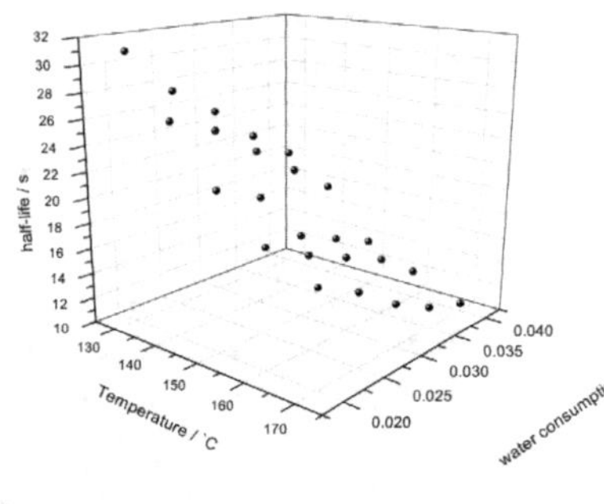

Figure 7. Half-life of foamed asphalt.

It can be seen from Figure 6 that the expansion rate of different water consumption systems of the pitch increases with the rise of temperature at early stage, and then decreases with the rise of temperature, with a critical value of 160°C. The main reason for the increase of temperature before 160°C is that the increase of temperature accelerates the vaporization of water and promotes the formation of foamed asphalt[18]. When the temperature is too high, the asphalt will gradually move from non-Newtonian fluid into Newtonian fluid, resulting in the decrease of the surface tension of asphalt and the rapid loss of foam asphalt, resulting in poor effect of asphalt foaming. If the temperature exceeds 180°C, the asphalt may occur varying degrees aging, so the foaming temperature should be lower than 180°C. As shown in Figure 7, the half-life decreases with the increase of temperature, showing a rule different from the expansion rate. Therefore, the evaluation index of single expansion rate or half-life cannot be used to evaluate the effect of asphalt foaming comprehensively.

3.2. Effect of water consumption on foaming effect
According to Figure 6 and 7, that the expansion rate of asphalt and the half-life of asphalt show the opposite rule in the range of 2% to 4% water consumption. The expansion rate of asphalt increases with the temperature, and the increase is less than 18%. The half-life decreases with the increase of temperature, and the overall decrease is less than 20%. It proved that the increase of water consumption promotes the burst of foamed asphalt and reduces the stability of foamed asphalt.

3.3. Determination of foaming conditions
From the above, it can be concluded that after the volume expansion of asphalt, the thickness of asphalt film turns thinner, which decreases of asphalt stability and half-life. The two trends are opposite. Therefore, the intersection method analyze the best water consumption under two variable conditions more intuitively and accurately, it's adopted it to determine the expansion rate and half-life of asphalt foaming, and quantitatively analyze the best conditions of asphalt foaming. The change of expansion rate and half-life with water consumption is shown in Figure 8:

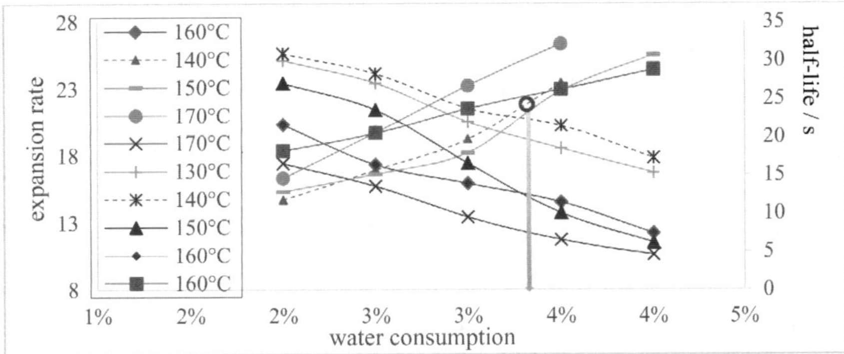

Figure 8. Relationship between water consumption, expansion rate and half-life.

According to the optimal design method, the curve intersection point data reflects the optimization. At the same temperature, when the half-life and the expansion rate reach the maximum at the same time, it is the optimal value. Therefore, the foaming effect of 140 ℃, 3.8% water consumption is the optimal choice, which is the best foaming condition for PBT / ABS resin reduced foam asphalt.

4. Conclusions
By altering the asphalt temperature and foaming water consumption in preparation of PBT / ABS resin foam asphalt, based on the existing foaming equipment and methods, the influence of water consumption and asphalt temperature on asphalt foaming is explored, and the optimum expansion parts are determined by expansion ratio, half-life and foaming index. The main conclusions are as follows:

(1) The expansion rate, half-life and foaming index effectively evaluate the foaming effect of asphalt. The expansion rate and half-life perform the opposite trend with the water consumption. Only by the foaming conditions with larger expansion rate and half-life achieve the best foaming conditions, it shows that the water consumption and asphalt temperature impacts on the foaming effect markedly.

(2) At the same temperature, the increase of water consumption increases the expansion rate of asphalt and decreases the half-life. When the water consumption is the same, the higher the temperature is, the greater the expansion rate is. But when the temperature reaches 160°C, the expansion rate does not increase with the increase of the asphalt temperature, and the asphalt with different temperature corresponds to different optimal foaming water consumption.

(3) considering the comprehensive expansion rate, half-life and foaming index, the optimum foaming conditions of the foamed asphalt are that the temperature of the asphalt is 140°C and the water consumption is 3.8%. The expansion rate of the foamed asphalt is 21.3 and the half-life is 22.5s.

Acknowledgments
The authors would like to thank 2018, 2019 Technology Program of Shanghai Municipal Engineering Design Institute (Group) Co., Ltd. No. K2018K081, K2019K063 for funding of this project, Tian Jianjun(SMEDI), Prof. Huang Weidong(Tongji University), and Asso. Prof. Yan Chuanqi(SWJTU) for helpful and constructive prophase studies.

References
[1] Asphalt Academy 2002 The design and use of foamed asphalt treated materials *Pretoria* 1
[2] Mallick RB Pelland R Hugo R 2005 Use of accelerated loading equipment for determination of long term moisture susceptibility of hot mix asphalt *Int J Pavement Eng June* **6(2)** 125–36
[3] Lv Weimin Shi Fang Zhi 2005 Research and application of cold recycling technology of foamed asphalt Road *construction machinery and construction mechanization* **22** 3
[4] A.R. Pasandín I Pérez 2014 Effect of ageing time on properties of hot-mix asphalt containing recycled concrete aggregates *Construction and Building Materials* **52** 284–293
[5] Martin Hugenera Manfred N. Partla and Markus Morant 2014 Cold asphalt recycling with 100% reclaimed asphalt pavement and vegetable oil-based rejuvenators *Road materials and pavement design* **15(2)** 239–258
[6] Milton L J and Earland M. G 1999 “Design guide and specification for structural maintenance of highway pavements by cold in-situ recycling” *Transportation Research Laboratory* 386
[7] Zhu F Liang R 2008 Developing simple lab test to evaluate HMA resistance to moisture, rutting, thermal cracking distress. University of Akron
[8] Huang Ming Wen Xuejun Wang Linbing 2017 Influence of foaming effect, operation time and health preserving properties of foam epoxy asphalt mixtures *Construction and Building Materials* **151** 931–938
[9] JTG E20 Beijing 2011 Standard test method of bitumen and bituminous mixtures for highway engineering China Communications Press
[10] Puppala An J. Hoyos et al 2011 Resilient moduli response of moderately cement-treated reclaimed asphalt pavement aggregates *Journal of Materials in Civil Engineering* **23** 990
[11] Marquis B R.L. Bradbury S Colson R.B Malick Y.V Nanagiri J.S. Gould S O’Brien and M. Marshall Washington 2003 “Design, construction and early performance of foamed asphalt full depth reclaimed (FDR) pavement in maine.” DC: CD-ROM of TRB 82th Annual Meeting
[12] Xiang S Huang B Shrum E D et al 2012 Laboratory evaluation of moisture susceptibility of foamed warm mix asphalt containing high percentages of RAP *Construction & Building Materials* **35** 125
[13] Grilli, Andrea Graziani et al 2016 Volumetric properties and influence of water content on the compact ability of cold recycled mixtures *Materials & Structures* **49** 1
[14] Xiang Shu Baoshan Huang Emily D Shrum Xiaoyang Jia 2012. Laboratory evaluation of moisture susceptibility of foamed warm mix asphalt containing high percentages of RAP. *Construction and Building Materials* **35** 125–130
[15] Akeroyd F M and B J Hicks 1988. Foamed bitumen road recycling *Highways* **56(1993)** 42-45
[16] Yao Qizhong 2018 Microcosmic study and performance analysis of the strength forming structure of recycled foamed asphalt mixture Harbin Institute of Technology
[17] Chen X Huang B Evaluation of moisture damage in hot mix asphalt using simple performance and superpave indirect tensile tests *Construction and Building Materials* **22** (**2008**) 1950–62
[18] Van Wijk A and L.E. Wood Washington 1983 “Use of foamed asphalt in recycling of an asphalt pavement.” Transportation Research Record 911 DC: TRB, National Research Council 96-103.

Influence of freezing-thawing cycles on strength of fiber recycled concrete column

Shihai Dong [1,*], Yuanyuan He [2], Jingqiang Huang [1] and Jiangfeng Dong [2]

[1] POWERCHINA Urban planning & design institute Co,. Ltd. Guangzhou 511458, China;
[2] College of architecture and environment, Sichuan university, Chengdu 610065, China

*Email: dongshihai204@163.com

Abstract. In order to reduce the influence of freezing-thawing cycle on the recycled concrete and improve the use of recycled concrete in cold regions, this paper systematically explored the influence of different variables such as the mix ratio of recycled aggregate, the content of basalt fiber and freezing-thawing cycle on the mechanical properties of recycled concrete.The results show that adding basalt wire can improve the bearing capacity and ductility of the recycled concrete, but the mechanical properties of the recycled concrete are greatly reduced after the freeze-thaw cycle.

1. Introduction

According to relevant data analysis, the proportion of construction waste in urban garbage is up to 40%, and the proportion of waste concrete blocks is up to over 60% [1-2].At present, for the recycle of waste concrete has become a major attention focus in the construction industry and the environmental protection industry, but the recycled concrete because of the high porosity, water imbibition is strong, such as large shrinkage creep characteristics [3-5], cause its bearing capacity and durability under the action of freeze-thaw cycle, decrease [6-8], the serious influence the use of recycled concrete in cold region.Therefore, scholars have done a lot of research on improving the freeze-thaw performance of recycled concrete. The relevant research results show that the replacement rate of recycled aggregate has a great impact on the bearing capacity and durability of recycled concrete.The high porosity makes the recycled aggregate more affected by freeze-thaw than the original concrete.The mechanical properties of recycled concrete are also affected by the water-cement ratio, and the resistance of recycled aggregate to freeze-thaw cycle is higher than that of primary aggregate under the same water-cement ratio [9-11].In terms of the porosity of recycled aggregate, studies have shown that the fiber reinforcement technology can effectively improve the carrying capacity by bridging internal voidage, but the degree of improvement in mechanical properties varies with the fiber types and fiber content [6-7].

In this study, the influence of freezing-thawing cycle on the mechanical energy of specimens was analyzed by adding basalt fiber and the proportion of recycled aggregate, and an optimization scheme was proposed to optimize the freezing-thawing resistance of recycled concrete.

IOP Conf. Series: Materials Science and Engineering 1028 (2021) 012011 doi:10.1088/1757-899X/1028/1/012011

Table 1. Coarse aggregate material characteristics.

Aggregate category	Apparent density (kg/m^3)	Bulk density (kg/m^3)	Moisture content (%)	Crushing index (%)
Native aggregate	2347	1382	0.25	11.58
Recycled aggregate	2172	1219	2.25	13.42

Table 2. Proportion of recycled concrete.

Substitution rate(%)	Cement	Water	Gravel	Native aggregate	Recycled aggregate
0	1	0.53	1.91	3.55	0
50	1	0.53	1.86	1.72	1.72
100	1	0.53	1.86	0	3.55

2. Experimental set-up

2.1. Materials and specimens

The test specimen is a square short column, of which the square short column size is 100mm 100mm 400mm.The primary aggregate of Concrete used for pouring was screened according to Technical Requirments and Test Method of Gravel and Crushed Stone for Ordinary Concrete (JGJ53-92), and the aggregate with particle size of 2.36~19mm was taken.The basic performance index of sand crushed stone (or gravel) for ordinary concrete was measured according to Standard for technical requirements and test method of sand crushed stone (or gravel) for ordinary concrete (JGJ-2006).The recycled aggregate is derived from the waste materials of buildings after the earthquake in dujiangyan, sichuan province. It is made by manual screening after the aggregate crusher is broken mechanically. The particle size range is 4.75 ~ 19mm.The fine aggregate was river sand, the apparent density was 1460kg/m, the sediment content was 1.34%, the fineness modulus was 2.8, and the water was tap water.At the same time, the concrete mixture was mixed with basalt fiber with the volume content of 2kg/m, the fiber length was 15~19mm and the diameter was 13 mm.Due to the large water absorption rate of recycled aggregate, the water absorption rate of recycled aggregate was determined to be 2.25% after water absorption and saturation treatment before mixing and pouring concrete.The relevant indexes of recycled aggregate and the mix ratio of experimental concrete are shown in Table 1 and Table 2.

In order to determine the mechanical properties of recycled concrete with different replacement rates of recycled concrete aggregate, the configured recycled concrete with different replacement rates was cast into a 150mm×150mm×150mm cube test block to determine its compressive strength and splitting tensile strength. The results are shown in Table 3.

Table 3. Mechanical parameters of recycled concrete.

Substitution rate(%)	Compressive strength(MPa)	Axial compressive strength(MPa)	Splitting tensile strength(MPa)	Elasticity modulus(MPa)	Poisson ratio
0	34.8	25.2	2.7	22.1	0.22
50	33.2	23.5	1.4	25.1	0.23
100	28.7	23.1	1.5	20.1	0.24

2.2. Experiment scheme

A total of 27 specimens of short square columns of recycled concrete were involved in this experiment, which were divided into 3 groups. Each group was composed of 3 specimens each with the substitution rate of recycled aggregate of 0%, 50% and 100%, and the tail Numbers 1, 2 and 3 were used respectively.The B0 group was the basic group, and the influence of the substitution rate of recycled aggregate on the concrete performance was studied.Group B1 is a concrete specimen group with the content of basalt fiber filament of 2kg/m, which is used to compare the improvement degree of basalt fiber filament on the bearing capacity of recycled aggregate.B2 is the freezing-thawing cycle group of fiber regenerated concrete, which is used to compare the influence of freezing-thawing cycle on the performance of fiber regenerated concrete.

2.3. Loading and freeze-thaw scheme

2.3.1. loading scheme. The static load test was carried out by a 200T press, and the loading scheme of each specimen was the same: the specimen was preloaded before loading, and the method of graded loading was adopted. The loading speed was 0.5kN/s, and the loading rate was 5kN per stage. The load was maintained for 3min.The specific load-time relationship is shown in Figure 1.

2.3.2. Freeze-thaw scheme. The freezing-thawing scheme of this experiment is conducted in strict accordance with the "fast freezing method" for frost resistance performance test in Standard for test method of long-term performance and durability of ordinary concrete (GB/T 50082-2009).Because of the early trials found that plain concrete short column specimen manager after 55 times freeze-thaw cycle has been severely damaged, so the test set the freeze-thaw cycles to 40 times. Each time the cycle time of freezing and thawing is 4 h, used for the melting time accounts for a quarter. In the process of freezing and melting, specimen center temperature control in -18 ± 2 °C and 15 ±2°C range, temperature is not higher than 17 °C specimen center at any time and not less than -20 °C.The temperature-time relationship between the specific freeze-thaw liquid and the specimen is shown in Figure 2.

2.3.3. Deformation and displacement monitoring. Vertical axial and annular strain gauges were arranged from the top of the column to the bottom of the column at 50mm, 200mm and 350mm, respectively. Strain data were collected every 10kN during the loading process to monitor the vertical and annular strain of the specimen.Three displacement meters were used to contact the three angles of the upper part of the specimen to measure the vertical displacement of each Angle during the loading process.

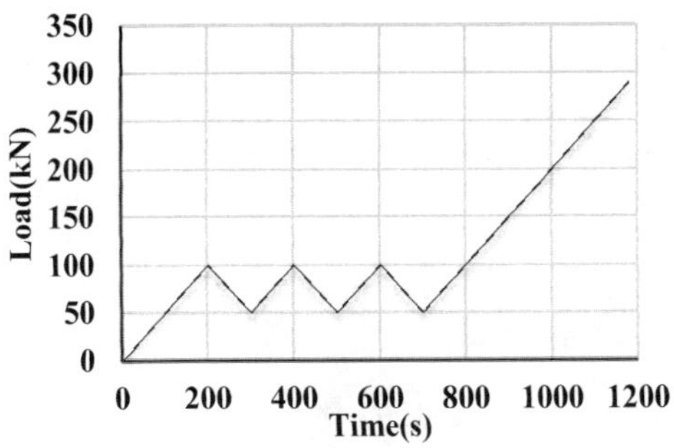

Figure 1. Load time curve.

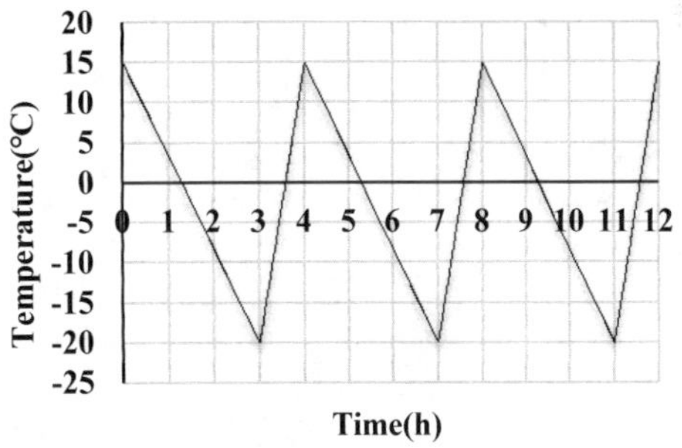

Figure 2. Freeze-thaw temperature curve.

3. Results and analysis

3.1. Failure pattern of specimen

3.1.1. Non-freeze-thaw specimen. The change of concrete short column mainly presents three stages: at the initial stage of loading, the load-strain curve of the specimen presents linear transformation, and the specimen surface has no obvious change; With the increase of load, the strain and vertical displacement of concrete gradually increase, and the transverse strain gradually appears in short columns. When the load increases to 75-85% of the ultimate load, the specimens reach the ultimate tensile strain of concrete due to the large transverse deformation and start to appear tiny cracks, accompanied by the sound of clear concrete splitting. As the load continues to increase, the width of cracks on the column surface increases rapidly and the number of cracks increases, and the cracks continue to expand to both ends. After that, the crack width and displacement of the test short column increase sharply, and the strain gauge value at each measuring point changes sharply or overflows, while the load increases slowly. When the load reaches the limit load, the concrete crack penetrates and the specimen is destroyed. The failure modes of concrete short columns are mainly axial compression failure, splitting failure and baroclinic failure. The failure patterns of the non-freeze-thaw specimens are shown in Figure 3.

Figure 3. Failure pattern of non-freeze-thaw specimens.

3.1.2. Freeze-thaw specimen. During the loading process of the specimen, there was no obvious phenomenon on the surface of the specimen at the early stage of loading, and when the load increased to a certain extent, the specimen suffered from brittle failure.As can be seen from the figure, after freeze-thaw, the specimen mainly presents the form of splitting failure, and the damage degree of the specimen is relatively serious when it reaches the limit load.The square column is mainly crushed by axial concrete, and the failure position of short concrete column is mainly in the middle and upper end of the column.Compared with the non-freezing-thawing specimens, the concrete fragments were more crushed after the specimens were damaged after freezing-thawing. The specimens were obviously broken into several parts, and there was an obvious separation between aggregate and concrete in the specimens.The failure forms are similar and the different replacement rates of recycled aggregate have no obvious effect on the failure modes of concrete.The failure patterns of freeze-thaw specimens are shown in Figure 4.

Figure 4. Failure pattern of freeze-thaw specimen.

Table 4. Summary of test results.

Specimen number	Ultimate load(kN)	Displacement(mm)	Failure modes
B01	195.69	1.77	Axial failure
B02	133.71	1.74	Diagonal-compression failure
B03	116.75	1.65	Diagonal-compression failure
B11	220.80	3.54	Splitting failure
B12	214.37	2.34	Axial failure
B13	170.41	1.76	Axial failure
B21	158.25	1.36	Splitting failure
B22	124.53	1.43	Splitting failure
B23	105.42	1.46	Splitting failure

3.2. Bearing capacity analysis

The ultimate bearing capacity and ultimate displacement of each specimen are shown in Table 4. The Load-Displacements of each specimen are shown in Figure 5-6.

According to the data in Table 4, the bearing capacity of the specimens B02 and B03 decreased by 31.6% and 40.3% respectively, indicating that the bearing capacity of the specimens gradually decreased with the increase of the replacement rate of the restrained specimens.The bearing capacity of the specimens in group B1 increased by 12.8%, 60.3% and 45.9%, respectively, compared with that in group B0. Among them, the best effect was achieved when the replacement rate was 50% and fiber was added. The bearing capacity of the specimens was fully exerted when the failure type was axial compression failure, while the bearing capacity of the specimens was lower when the failure type was baroclinal failure.Fiber recycled concrete short column after 40 weeks time of freezing and thawing bearing capacity is greatly reduced, B2 specimen relative B1 group specimens bearing capacity to reduce the amplitude decreased respectively 28.3%, 41.9%, 38.1% respectively, the replacement rate of 50% reduction of maximum bearing capacity, the specimens showed that the replacement rate of concrete resistance to freezing and thawing cycle ability is low.The results show that the resistance of recycled aggregate to freezing-thawing cycle is worse than that of primary aggregate, but the suitable mix ratio of recycled aggregate is beneficial to the structural performance of recycled concrete.

From the data in Figure 5, it can be seen that the replacement rate of recycled aggregate has little impact on the ductility of concrete.With the addition of fiber, the ultimate displacement of concrete increased significantly, and the ultimate displacement of specimens with the substitution rate of "0%", "50%" and "100%" increased by 100%, 34.5% and 6.67%, respectively. The load-displacement trend of specimens with the substitution rate of 100 was the steepest, and the strength was the lowest, which was also the earliest to enter the plastic stage. It can be seen from Figure 6 that the performance of recycled concrete has a great change after freeze-thaw. Under different substitution rates, the ultimate displacement decreases by 61.5%, 38.9% and 17.05% respectively under the corresponding substitution rates.The specimen with the substitution rate of "0" develops the most slowly in displacement and the elastic stage is longer, followed by the specimen with the substitution rate of "50" and the substitution rate of "100" is the fastest in displacement development. The higher the substitution rate, the lower the strength of the concrete short column.

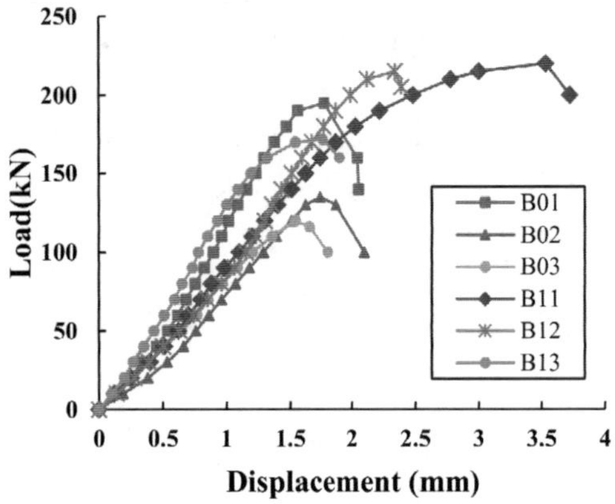

Figure 5. Load-displacement curve of non-freeze-thaw specimen.

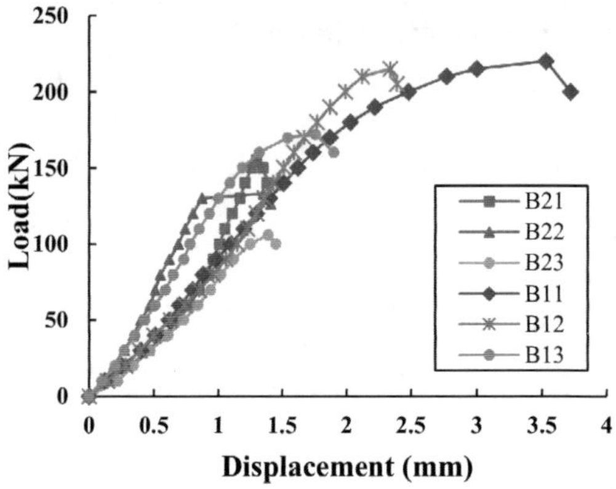

Figure 6. Load-displacement curve of freeze-thaw specimen.

4. Conclusion

The test results show that the bearing capacity of recycled concrete can be improved by adding basalt fiber, and the bearing capacity can be increased by up to 60%, and the ductility of recycled concrete can be improved greatly. Under the influence of freezing-thawing cycle environment, the bearing capacity of recycled concrete decreased significantly, and the specimen with the replacement rate of 50% even decreased as much as 41.9%. The reduction rate of ultimate displacement after freezing-thawing decreased with the increase of the replacement rate of recycled aggregate.

References

[1] Xiao Jiangzhuang 2008 Recycled concrete[M] *China building industry press*
[2] Xiao. J, Li W, Fan Y 2012 An overview of study on recycled aggregate concrete in China(1996-2011) *Construction and building materials* **31(1)** 364
[3] Wijayasundara, Mayuri, Mendis, Priyan, Crawford, Robert H 2017 Methodology for the integrated assessment on the use of recycled concrete aggregate replacing natural aggregate in structural concrete *Journal of cleaner production* **166(10)** 324
[4] Omary S, Ghorbel E, George Wardeh 2016 *Construction and building materials* **108(1)** 163
[5] Zheng Yuanxun, Yang Peibing, Kang Haigui 2016 *Journal of Zhengzhou university (engineering science)* **37 (5)** 27
[6] Medina C S, Rojas M I, Fras M 2013 Freeze-thaw durability of recycled concrete containing ceramic aggregate *Journal of cleaner production* **40(2)** 151
[7] Xiao Jianzhuang, Wang Changqing, Sun Zhenping 2016 *Journal of civil engineering* **48 (2)** 9
[8] Yu H F, Ma H X , Yan K 2017 *Construction and building materials* **137(1)** 104
[9] Duan A, Jin W L, Qian J R 2011 Effect of freeze-thaw cycles on the stress-strain curves of unconfined and confined concrete *Materials and structures* **44(1)** 1309
[10] Qin X C, Meng S B, Cao D F 2016 Evaluation of freeze-thaw damage on concrete material and prestressed concrete specimens *Constr build mater* **125(1)** 892
[11] Hong Lei, Zhang Yujian, Wang Suyan 2016 *Journal of water resources and architectural engineering* **014 (03)** 200

Research on influencing factors of high temperature performance of new type noise reduction micro-surfacing

Xinrui Li[1,*]

[1] Road engineer, Shanghai municipal engineering design institute (group) co., ltd., Shanghai 200092, China

*Email: lixinrui@smedi.com

Abstract. The micro-surfacing has received widespread attention with its excellent anti-skid performance, waterproof performance, construction characteristics and significant cost advantages. However, the problem of the conventional micro surfacing high driving noise has been criticized. The new noise reduction micro-surfacing has appeared to greatly reduce the traffic noise to a considerable extent and solve the main bottleneck of the further promotion to the micro-surfacing. The noise reduction micro-surfacing proposed in this study has a special binder and gradation, as well as a specific rubber powder filler. Guided by the change rate of the wheel rut width, focused on the four key variables of oil-stone ratio, cement, mineral powder and rubber powder content, the high-temperature performance of the noise reduction micro-surfacing mixture is studied. The research shows: based on the special noise reduction gradation proposed in this article, in order to obtain a new type of noise reduction micro-surfacing with the best anti-rutting performance, the best oil-stone ratio is 10.5%, the best cement content is 1%, and the rubber powder should not exceed 2%.

1. Introduction

The highways and urban roads that have been opened to traffic in China have entered the peak period of maintenance on a large scale. The maintenance of asphalt pavements of various levels of highways and urban roads has received more and more attention. As a maintenance method with excellent cost-effectiveness and wide practicability, the micro-surfacing area is favored by many engineers [1]. For a long time, the noise caused by vehicles walking on the road surface of the micro-surfacing is higher than that of the ordinary road surface, which troubles the researchers and users. The birth of the noise reduction micro-surfacing （it only reduces the noise on the basis of conventional micro surfacing） after grading and composition reconstruction has brought a new look to this maintenance measure, but due to the adjustment of the gradation and the replacement of the binder, the increase of the asphalt-aggregate ratio will inevitably impact the density and the reduction of the clearance rate on the internal structure of the micro-surfacing, impact resistance to shear stress, among them, the most critical impact on the road performance is the high temperature performance. Considering that the change in the amount of coarse aggregate in the gradation will affect the high temperature performance of the mixture [2-3], the influence factors of the high temperature performance of the noise reduction micro-surfacing are studied, in order to provide an important reference for the design of the mix ratio composition of the noise reduction micro-surfacing during construction.

2. Experimental

2.1. Material

In order to make the research object representative, the wide-area type EM520 slow-cracking quick-setting emulsifier was used, which has a high tolerance for aggregate types and processing levels [4]. The technical indicators of modified emulsified asphalt emulsified by EM520 are shown in Table 1.

Table 1. Modified emulsified asphalt for micro surfacing.

Test items		Unit	Result	Standard requirement	Test method
sieve residue（1.18mm）		%	0.04	≤0.1	T 0652
charge			+	+	T 0653
standard viscosity C25,3		S	19.2	12～60	T 0621
asphalt evaporation residue margin		%	64.2	≥60	T 0651
Properties of asphalt evaporation residue	penetration（25℃）	0.1mm	70.1	40-100	T 0604
	softening point（R/B）	℃	64.3	≥53	T 0606
	ductility（5℃）	cm	35.1	≥20	T 0605
Storage stability	1d	%	0.41	≤1	T 0655
	5d	%	2.53	≤5	

The coarse aggregates used in the test are basalt produced in Jiangsu and the fine aggregates are limestone produced in Jiangsu. According to the noise reduction grading design, the aggregates are sieved into 0-3, 3-5 and 5-10 grades for the synthesis of the target gradation. Mineral powder will not participate in the noise reduction micro-surfacing gradation design, but it will be used as an external additive. The screening results of aggregate are shown in Table 2.

Table 2. Quality requirements for coarse and fine aggregates at low noise micro-surfacing.

Material name	project	requirement	Test method
Coarse aggregates	crushing value（%）, ≤	28	T 0316
	Los Angeles abrasion loss（%）,≤	30	T 0317
	burnish value ≥（BPN）	42	T 0321
	sturdiness≤（%）	12	T 0314
	acicular content≤（%）	15	T 0312
Fine aggregates	sturdiness≤（%）	12	T 0340
Mineral	sand equivalent≥（%）	60	T 0334

Cement is added as a filler into the mixture, should be dry, loose, without agglomeration, and should meet the relevant requirements in the "Technical Specification for Construction of Highway Asphalt Pavements" (JTG F40). The amount of cement added must be determined through a mixture design test.

Table 3. Technical requirements for physical properties of rubber powder.

Technical index	Measured value	quality requirement
Fiber content（%）, ≤	0.3	1
Metal content（%）, ≤	0	0.05
Relative density	1.17	1.1～1.3
Other impurities content（%）	0	0
Acetone extract, ≤	10	22
Carbon black, ≥	40	28
Rubber hydrocarbons, ≥	51	42

The low noise effect of the micro-surfacing mixture used in this study mainly comes from two aspects, the special gradation design and rubber powder. The addition of appropriate rubber is one of the key elements to reduce noise [5]. The rubber powder should be made by normal temperature grinding or granulation method [6]. The technical indicators and measured values of rubber powder used in the research are shown in Table 3.

2.2. *Grading*

Aggregate grading has an important impact on the road performance of the micro-surfacing, and it is also another key to driving noise. Appropriately reducing the use of coarse aggregate, adding the medium-grain aggregate (3-5mm) with higher Los Angeles abrasion value, strengthening the squeezing effect between aggregates, reducing the difference between the laying thickness and the maximum particle size of the aggregate, using discontinuous grading to increase the proportion of coarse aggregate are the design principle of noise reduction gradation. However, the noise reduction treatment from the gradation point of view has an impact on the road performance, and the asphalt-aggregate ratio decreases with the increase of the coarse aggregate, which will lead to a decrease in durability. Increasing the asphalt-aggregate ratio helps to improve the durability of the micro-surfacing, but it will cause oil flooding and sticking to the surface structure of the micro-surfacing [7]. Table 4 shows the gradation of noise reduction micro-surfacing.

Table 4. noise reduction gradation.

Mesh size /mm	9.5	4.75	2.36	1.18	0.6	0.3	0.15	0.075
Passing rate /%	100.0	85.5	60.2	36.4	28.9	18.8	15.2	7.0

2.3. *Test methods*

According to the requirements of the performance indicators at the micro-surfacing in the construction technical specifications, the rutting deformation test is carried out in accordance with the designated test T0756 in the " Standard Test Methods of Bitumen and Bituminous Mixtures for Highway Engineering " (JTG E20-2011) .

3. Rutting resistance

3.1. *Asphalt-aggregate ratio*

The test uses the gradation of noise reduction micro-surfacing from 9.0% to 12.0% of the asphalt-aggregate ratio. Making the test pieces according to test specifications, used the load wheel tester to simulate the wheel rolling on the mixed material at the micro-surfacing after forming, after the specified number of actions, the rutting depth and width deformation of the sample were measured, and the rutting depth per unit thickness of the sample and the lateral deformation per unit width were used to evaluate the anti-rutting ability of the mixture. The test results are shown in Table 5.

Table 5. Deformation rate of wheel track width under different asphalt-aggregate ratio.

Gradation of noise reduction micro-surfacing	Asphalt-aggregate ratio /%						
	9.0	9.5	10.0	10.5	11.0	11.5	12.0
Deformation rate of wheel track width /%	8.02	7.33	6.45	4.57	4.83	5.48	6.52

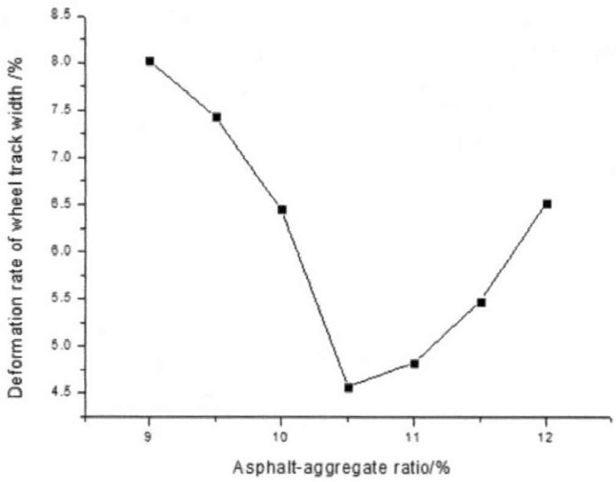

Figure 1. Variation curve of wheel rut width with different asphalt-aggregate ratio.

For the noise reduction micro-surfacing mixture proposed by the new design method, the difference from the conventional micro-surfacing is that with the increase of the asphalt-aggregate ratio, the change rate of the wheel track width decreases first and then increases. This is reflected in the test results in Figure 1. There is an optimal asphalt-aggregate ratio to optimize the anti-rutting ability of the mixture at the low noise micro-surfacing. Based on the test results, this study recommends 10.5% as the best asphalt-aggregate ratio.

Generally, for hot-mix asphalt, the anti-rutting ability has a great relationship with factors such as mineral clearance ratio, asphalt-aggregate ratio, and mixture void ratio. The rut depth will gradually decrease with the increase of the void ratio of the mixture, and the void ratio is the main factor affecting the rutting resistance. However, due to the thin paving thickness at the micro-surfacing, it is mostly resistant to moderate and mild rutting. It is not enough to simply adjust the gradation to improve the high temperature performance. The appropriate material composition should be comprehensively determined by adjusting the asphalt-aggregate ratio, changing the types of additives, and combining actual engineering experience and construction environment conditions to improve the high temperature performance of the micro-surfacing mixture. The recommended value of 110.5% of the best asphalt-aggregate ratio here can be used as the initial value of the design and construction mix ratio at the noise reduction micro-surfacing.

3.2. Cement and mineral powder

Selected the special gradation to determine the best asphalt-aggregate ratio (10.5%) and the corresponding best content (2.0%), added different proportions of cement and mineral powder to conduct wheel rutting test to study the effect of additive types and content on high temperature performance. The test results are shown in Table 6 and Table 7.

Table 6. Wheel rutting test resu0lts under different cement content.

Cement content /%	0	0.5	1	1.5	2
Deformation rate of wheel track width /%	4.57	4.01	2.67	2.53	2.49

Table 7. Wheel rutting test results under different mineral powder content.

Mineral powder content /%	0	1	2	3	4
Deformation rate of wheel track width /%	4.57	4.32	3.79	3.71	3.53

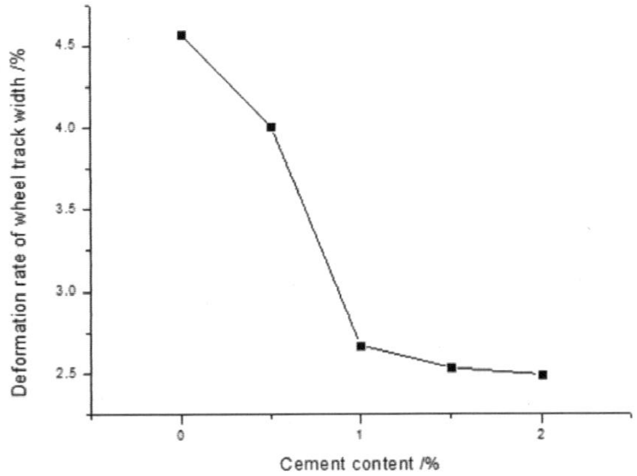

Figure 2. Influence of cement content on the high temperature performance of micro-surfacing.

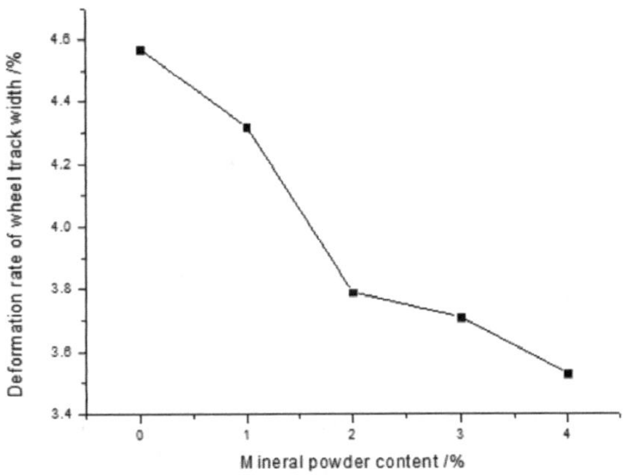

Figure 3. Influence of mineral powder content on the high temperature performance of micro-surfacing.

The curve in Figure 2 shows that the addition of a small amount of cement can effectively improve the rutting resistance of the micro-surfacing. When the addition amount is 0.5% and 1.0%, the change rate of the rut width of the mixture at the micro-surfacing is 12.3% and 41.5% higher than that of no cement. However, when the cement content is greater than 1.0%, the change rate of the wheel track width tends to be stable and no longer changes with the increase of the cement content. It can be considered that for the high temperature performance of graded noise reduction micro-surfacing, 1.0% cement content is the best value. Too much cement cannot further improve the ability of the mixture to resist rutting deformation at the micro-surfacing.

It can be seen from Figure 3 that the addition of mineral powder does not significantly improve the rutting resistance of the micro-surfacing. When the mixing amount of mineral powder is 1% and 2%, the wheel rut width deformation rate of the mixture is only increased by 5.5% and 17.1% respectively compared with that without mineral powder. After that, adding mineral powder has no significant effect on improving the anti-rutting ability of the micro-surfacing. The reason is that the addition of mineral powder changes the gradation of the mixture and increases the number of fine aggregates. Although reducing the porosity of the mixture can improve rutting resistance, the mineral powder will absorb more modified emulsified asphalt. If the asphalt-aggregate ratio is not increased, the mixing property at the micro-surfacing will be reduced to a certain extent, the aggregate cannot be fully coated by the asphalt, and the ability of the mixture to resist rutting after emulsification will be weakened. In the whole, it is not significant to increase the amount of mineral powder alone to improve the high temperature performance of the micro-surfacing.

3.3. Rubber powder

Rubber powder has the effect of absorbing and weakening noise. Its addition can make the mixture at the micro-surfacing further reduce noise after paving is completed, and appropriately increase the elastic modulus of the micro-surfacing. The noise reduction gradation is still used, the best asphalt-aggregate ratio (10.5%) determined in the previous section is used, and the water content (2.0%) is mixed with rubber powder at intervals of 0.5% and 1.0% by mass to carry out wheel rutting deformation. Experiment to study the effect of different amounts of rubber powder on the high temperature performance of the micro-surfacing. The test results are shown in Figure 4.

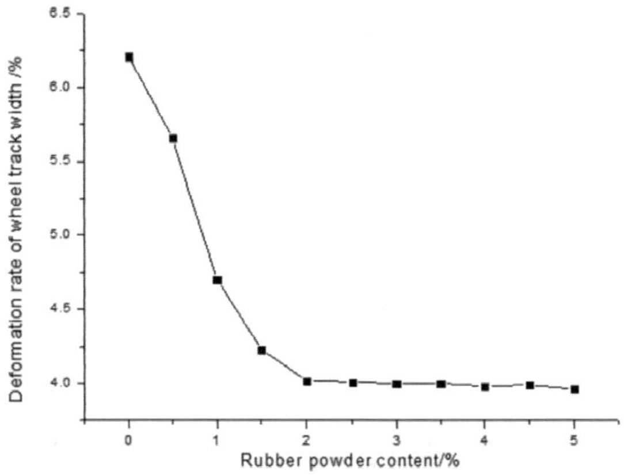

Figure 4. Change curve of wet wheel wear value with rubber powder content.

It can be seen from Figure 4 that adding rubber powder can also improve the high temperature performance of the micro-surfacing. 2.0% of rubber powder is a critical value. Between 0-2.0%, the high temperature performance increases significantly. After the critical range of 2.0%, the high temperature performance tends to be stable. More rubber powder cannot bring better road performance. Therefore, from the perspective of anti-rutting, 2.0% is the standard addition of rubber powder.

4. Conclusions

Based on the proprietary gradation proposed in this article, according to factors such as asphalt-aggregate ratio, cement, mineral powder and rubber powder, the effects of them on the high temperature performance of noise reduction micro-surfacing mixtures were studied. The main conclusions are as follows.

(1) On the basis of the noise reduction micro-surfacing gradation proposed in this article, by adjusting the ratio of oil-stone, the amount of water, the content of cement, mineral powder and rubber powder, a micro-surfacing mixture with better high-temperature performance can be produced.

(2) With the increase of the asphalt-aggregate ratio, the rate of change of the rut width at the noise reduction micro-surfacing decreases first and then increases. There is an optimal asphalt-aggregate ratio to optimize the rutting resistance of the mixture at the micro-surfacing. It is recommended to use 10.5% as the best asphalt-aggregate ratio at the noise reduction micro-surfacing, which can be used as the initial design value of the asphalt-aggregate ratio in actual construction.

(3) A small amount of cement can effectively improve the rutting resistance of the micro-surfacing, but when the cement content exceeds 1.0%, the lifting effect tends to be stable. The addition of mineral powder can also improve the rutting resistance of the micro-surfacing, but the affection is not obvious, because the mineral powder will absorb more modified emulsified asphalt. If the asphalt-

aggregate ratio is not increased, the aggregate will not be fully coated by the asphalt. The mixture will be demulsified, and the ability to resist rutting performance will be weakened.

(4) Adding rubber powder can improve the high-temperature stability of the micro-surfacing. According to the content and high-temperature performance curve, the performance tends to be stable after the rubber powder content exceeds 2.0%. From the perspective of anti-rutting, it is recommended 2.0% is the standard rubber powder addition amount for the noise reduction micro-surfacing.

References

[1] Liu Yuheng 2016 Research on the key technology of low-noise micro-surfacing processing technology Jiangsu: Master thesis of southeast university
[2] Zhong Jianchao He Zhiyong Yao Ailing et al. 2014 Optimization design of mixture grading at low noise micro-surfacing highway and transportation *Science and technology* **31** 6
[3] Son H J Kim Y J Baek JE et al. 2015 Field performance evaluation of micro- surfacing method and polymer slurry seal method used in national highway International *journal of highway engineering* **17** 1
[4] Mi Haichen 2001 Research on asphalt emulsifier and its application technology shanxi: Master thesis of Chang'an university
[5] Zhang Ran 2011 Research on the mechanism and noise reduction technology of pavement noise at the micro-surfacing Heilongjiang: Master thesis of Harbin institute of technology
[6] Wang Longfei Lin Jiangtao Fan Liang et al 2011 Discussion on the wet wheel abrasion test method of the mixture at the micro-surfacing *Petroleum asphalt* **35** 4
[7] Ju Hao Huang Xiaoming 2007 Research on influencing factors of mixture performance at micro-surfacing *highway* **43** 7

Study on non-destructive reinforcing material of modified one component polyurethane for tunnel crack disease prevention

Zhiwei Wang*, Weibin Ma, Zhaohui Ma

China railway research institute group co., ltd. railway construction institute, Beijing, 100081, China

*Email: 240229397@qq.com

Abstract. In view of the defects of tunnel lining structure such as cracking and block dropping, cavity behind, honeycomb and pitted surface, construction joint dislocation and other defects, the technical research direction for non-destructive reinforcement materials in tunnel is proposed. The polyamide curing coating agent and one-component polyurethane protective material with high adhesion, good elongation, large tensile strength and rapid drying are developed. The above materials are described The application technology and key points of brush coating, rolling coating and scraping coating can provide reference for related engineering application.

1. Introduction

China is the country with the longest railway operation mileage and the largest scale under construction in the world. By the end of 2019, the total length of China's railway operating tunnels is 18041 km, with a total of more than 16084[1]. According to the medium and long term railway network planning issued by the national development and Reform Commission, the railway operation mileage will reach 200000 km by 2030.

After the tunnel is in operation, the lining structure will have various defects or diseases due to unreasonable design, improper construction, geological disasters, operation years, climatic conditions and other reasons, mainly manifested as: lining cracking and block dropping, lining back cavity, honeycomb pitted surface, lining construction joint dislocation and other defects.

At present, the treatment measures for lining deterioration variation at home and abroad mainly include chiseling, caulking, cracking and grouting, all kinds of protective plates (NETs) and lining (arch), anchoring, local reconstruction, etc.[2-5]. For sections with serious diseases or insufficient lining thickness and strength, it is easy to cause secondary man-made damage in the process of reinforcement and treatment, which will damage the bearing capacity of the structure. The maintenance of force has a negative effect. In addition, for the existing operation lines, the above countermeasures and technologies are limited by the maintenance time. Therefore, the technical and economic problems involved are worth considering.

In recent years, non-destructive spray reinforcement technology has been widely used in the field of industrial and civil construction. By coating on the surface of concrete cracks and spalling areas, this kind of material can effectively strengthen the concrete in a certain range around the defects. This kind of technology concept is advanced and easy to implement, and has not been applied in the field of domestic railway engineering at present, and the related scientific research achievements are blank.

In this paper, a kind of high strength and high adhesion non-destructive reinforcement material for tunnel protection is proposed. Based on the demand analysis, the research and development technology and construction technology of the material are described in detail, which can provide some reference for engineering application.

2. Demand analysis

2.1. Types and manifestations of tunnel diseases
The forms of tunnel lining cracks are as follows: circumferential cracks, longitudinal cracks, oblique cracks, crescent shaped cracks, cross shaped cracks, etc.; the manifestations of lining block falling are: concrete pitted surface spalling, construction joint laitance falling block, etc. According to statistics, the crack disease of railway tunnel lining accounts for no less than 30%[3].

2.2. Construction characteristics of disease reinforcement project in tunnel
①The span of tunnel working face is large and the clearance area is large. The excavation span of single and double track tunnels of high-speed railway and single and double track tunnels of ordinary speed railway ranges from 6.98m to 13.30m, and the clearance is from 42.06 m2 to 100.00 m2. In strict compliance with the rules and regulations of railway tunnel construction, according to the characteristics of different tunnel span and tunnel operation space, select suitable and suitable rapid lifting equipment for operation, so as to improve the construction efficiency under the premise of ensuring safety.

②There are many facilities in the tunnel and the protection requirements are high. Tunnel equipment and facilities include catenary, contact suspension, support device, traction device, positioning pillar, communication and signal facilities, wall hanging cable and track rail, and most of the components are not allowed to touch.

③Single day construction time is short. The construction process is greatly affected by the maintenance time of skylight, and the vault is located above the electrified catenary, which can only be carried out after power failure with the cooperation of power supply department, which is closely related to the process arrangement, construction process, equipment performance, personnel quality and the cooperation degree of property owners.

2.3. Material selection and coating form
According to the analysis of tunnel lining structure characteristics, hydrological information, disease types, causes and manifestations, it is determined that the materials should have high strength, high adhesion and convenient construction. The developed materials are the combination of non-destructive reinforcing materials and protective coatings, i.e. epoxy sealing primer, modified polyurethane with high strength and high adhesion for tunnel protection.

Among them, the epoxy sealing primer has excellent permeability to the concrete substrate, which can penetrate into the micro holes and cracks on the concrete surface, adhere to the lining concrete and solidify, and bond the defective concrete into an organic whole after forming. The single component material should reach high adhesion and strength after modification design. If it is coated on the solidified primer coating, the two layers should have good adhesion and good flexibility, and will not cause interlayer damage and shedding due to slight dislocation, expansion and other deformation of the lining.

3. Research and development of one component polyurethane protective material

3.1. Introduction of one component polyurethane materials
Polyurethane (PU) is a kind of polymer with carbamate group repeating unit on the polymer chain. In 1937, Otto Bayer, a German chemist, first discovered that polyurethane (formula 3-1) could be prepared by the reaction of isocyanate with alcohol compounds, and then it was applied in industry.

Since the 1960s, China began to independently develop and develop polyurethane resin. In recent decades, it has been developed rapidly, with a wide range of products, and has been applied in all walks of life.

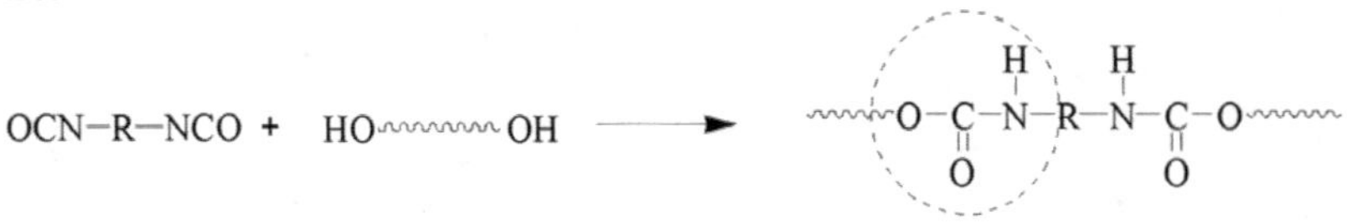
(1)

Polyurethane materials are widely used in Railway Engineering, such as polyurethane curing track bed, polyurethane sleeper, polyurethane elastic ballast cushion, polyurethane Micro Foam elastic cushion, etc.; the above polyurethane materials are also widely used in high-speed railway vehicles for heat preservation, noise reduction, bonding, sealing, insulation and other purposes[6].

At present, polyurethane materials are mostly composed of two components. It is necessary to adjust the amount of isocyanate and the ratio of hydroxyl or other functional groups in the two components before use. It is necessary to adjust the later curing state by adding catalyst or other additives, and then carry out construction after mixing evenly. The two-component polyurethane has only a short service life after mixing, which needs to be adjusted according to the site temperature. The process is complex and can not meet the requirements of short skylight time.

Compared with two-component polyurethane, one-component polyurethane is more convenient to use and does not need to be prepared later. When used, it can be crosslinked and cured by heating or other methods, so the operation is more convenient. The later stage of one component polyurethane coating is moisture curing, mainly through the reaction of NCO in polyurethane prepolymer with moisture in the air (formula 3-2), so it has a long service life. After use, the remaining materials can be sealed and stored to prevent cross-linking and curing.

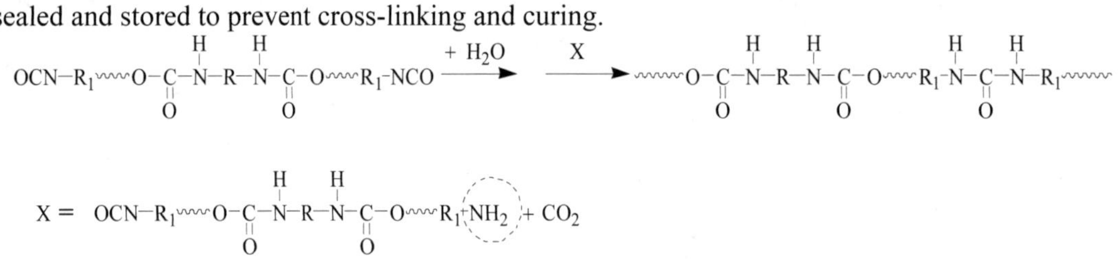
(2)

There are soft and hard segments in one component polyurethane. The main chain with good flexibility and polyol as soft segment, and isocyanate and chain extender as hard segment. Due to the thermodynamic incompatibility of hard and soft segments, the unique micro phase separation structure of polyurethane materials is caused. In addition to carbamate bond, polyurethane structure also contains ester bond, ether bond, urea bond, biuret bond, carbamate bond and acyl urea bond, which makes it easy to form hydrogen bond within and between molecules. There are many physical crosslinking points with strong binding force in linear polyurethane molecules. Therefore, it not only has the properties similar to amide groups, such as strength, wear resistance, oil resistance, but also has the heat resistance and solvent resistance of polyester, and the water resistance and flexibility of polyether.

In addition, isocyanate, the main raw material of polyurethane, is very active. It can not only combine with hydroxyl resin, but also with hydroxyl group in substrate to form strong chemical bond and hydrogen bond, which can enhance the adhesion with substrate.

3.2. Research on modification technology of one component polyurethane protective material

The common one component polyurethane protective material is easy to construct, but its strength is low after curing, the strength is 2 ~ 3Mpa, and the adhesion is poor. As a special material for tunnel protection, it should meet the characteristics of high strength, high adhesion, high elongation and fast surface drying speed, and should have good coating adaptability to ensure the coating quality and construction efficiency.

According to the above requirements, the material modification design was carried out

①To improve the adhesion performance, select the matching excellent epoxy sealing primer, improve the strengthening strength of concrete substrate, and enhance the interlayer adhesion with polyurethane topcoat.

②To improve the elongation and tensile strength, select suitable polyether diols, polyester diols and small molecular alcohols to synthesize prepolymers to meet the requirements of high elongation; increase the content of high cohesive groups, increase the chemical and physical crosslinking density, and increase the crystallinity to improve the strength of one component polyurethane materials.

③It is suggested that the surface drying time should be accelerated, and the non heavy metal catalyst should be selected to reduce the reaction activation energy and shorten the surface drying time.

④In order to meet the needs of spray coating, it is necessary to improve the sprayability, select the suitable solvent and reduce the viscosity of the system.

3.2.1. Epoxy sealing primer After curing, the epoxy resin has good mechanical properties, high strength, strong adhesion (including with concrete substrate), small shrinkage, good electrical insulation, excellent chemical resistance, and good stability of the resin itself. Moreover, the epoxy resin has good leveling and permeability on the concrete substrate, which can seal the pores and cracks on the surface of the base material, and improve the surface strength of the base material to a certain extent。 After curing, there are active functional groups on the surface of epoxy resin, which can participate in the reaction of other materials. For example - Oh can react with - NCO in polyurethane to improve the adhesion between the two materials. Bisphenol A epoxy resin is the most widely used epoxy resin. The curing reaction is generally ring opening cross-linking reaction between epoxy resin and amine curing agent (formula 3-3).

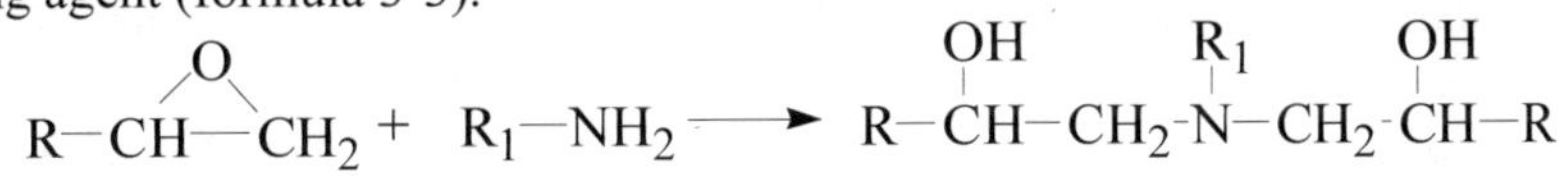
(3)

The curing strength and hardness of epoxy resin can be adjusted according to the epoxy equivalent and active hydrogen equivalent, and the compatibility of cured epoxy resin with polyurethane coating, such as interlayer adhesion and peel strength, should be met. High crosslinking density: the strength of epoxy resin is improved to a certain extent, but the wettability of polyurethane coating on the surface of epoxy resin becomes poor after too high, which makes the adhesion between layers and peel strength decrease. Low crosslinking density: the strength of epoxy resin decreases, so does the peeling strength. Therefore, it is necessary to select epoxy resin sealing primer with good compatibility, as shown in Table 1.

Table 1. Performance of self-developed multi type epoxy primer and matching table with polyurethane topcoat

General performance	Type I epoxy	Type II epoxy	Type III epoxy	Type IV epoxy
Surface drying time (h)	<5	<3	<2	<3
Actual drying time (h)	<24	<24	<8	<24
Adhesion of concrete substrate (MPa)	5.1	5.1	5.2	5
Adhesion with modified polyurethane (MPa)	4.6	4.3	3	≥5 Substrate Failure

According to the characteristics of polyurethane materials, the matching performance of primer and topcoat when different curing agents are used in epoxy primer is verified. The verification results are shown in Table 2:

Table 2. Verification results of matching performance of epoxy primer with different curing agents

General performance	Small molecular amine curing agent	Polyamide curing agent
Surface drying time (h)	<2	<3
Actual drying time (h)	<8	<24
Adhesion of concrete substrate (MPa)	≥5.2	≥5
Adhesion with modified polyurethane (MPa)	3	≥5 Substrate Failure

Comparison and analysis of the properties of the two kinds of curing agents:

Advantages of small molecular amine curing agent: high adhesion with concrete substrate, greater rigidity; fast curing of surface drying and solid drying; relatively low price. The disadvantages are: it is volatile and harmful to human body; after curing, the epoxy resin has high crosslinking density, low elongation and poor flexibility; the compatibility with flexible one-component polyurethane is poor, and the strength is only about 3Mpa.

Advantages of polyamide curing agent: high adhesion with concrete substrate; high adhesion with modified polyurethane layer, good compatibility between the two, strength greater than 5MPa; low volatility of polyamide curing agent. The disadvantages are: compared with small molecular amine curing agent, the dry time is slightly longer; the price is higher.

According to the comparative analysis of the performance test data of the above two kinds of curing agents, the comprehensive performance of epoxy sealing primer is better than that of small molecular amine curing agent when using polyamide curing agent. It has good compatibility with polyurethane topcoat, high strength, good flexibility and elongation. Therefore, "polyamide curing agent" coating agent is selected.

3.2.2. One component polyurethane coating material ① By comparing polyether diols 1000, 2000, 3000 and polyester diols 1000, 2000, 3000, the strength of single component polyurethane prepared from polyester diols is higher than that of polyether diol polyurethane, and the low temperature flexibility of polyether is better. With the increase of molecular weight of polyether and polyester diol, the elongation increases, but the acid-base resistance of polyester diol is poor.

② With the increase of multifunctional alcohols, the crosslinking density increases and the strength increases after curing, but the elongation decreases more obviously.

③ One component polyurethane was prepared by Aromatic Isocyanates (MDI, TDI, crude MDI). Its strength is higher than that of aliphatic isocyanate (hmdi, IPDI) polyurethane. The strength of crude MDI is the highest, but the stability is poor. The viscosity of MDI is high, which is not conducive to spraying. The activity of aliphatic isocyanate is low and the adhesion with epoxy primer is poor.

④ The solvent containing carbonyl group can form hydrogen bond with carbamate group and has better solubility. Therefore, ester solvent and ketone solvent have better viscosity reduction effect than other non-polar solvents.

⑤ Polyurethane catalysts include metal catalysts (organotin, organic bismuth) and amine catalysts. In order to shorten the surface drying time, amine catalysts with obvious catalytic NCO and water reaction effect and metal catalysts with obvious catalytic NCO and - Oh should be selected.

When the multi-functional raw materials and aromatic isocyanates are added, the crosslinking density of polyurethane topcoat is increased, and the high cohesive groups increase the strength of polyurethane coating, long-chain diols and polyols, and increase the elongation. Considering the later curing speed and elongation, the molecular weight of 2000-5000 diols and polyols were selected, and a certain amount of organic bismuth and tertiary amine catalyst were added to accelerate the crosslinking curing speed of polyurethane topcoat, so that the surface drying and real drying time were guaranteed.

Table 3 shows the performance comparison of "modified single group polyurethane material", "ordinary one component polyurethane waterproof material" and "two-component polyurea material". It can be seen that the "surface drying and real drying time" of modified single group polyurethane material is longer than that of polyurea material. A certain surface drying time also ensures that the material has enough wetting time on the concrete substrate, the crosslinking curing effect is guaranteed, and the reinforcement effect of lining concrete surface is improved.

Table 3. performance comparison of polyurethane protective coating, common polyurethane waterproof coating and two-component polyurea

Performance	Modified polyurethane materials	Ordinary polyurethane waterproof	Two component polyurea materials
Smoke toxicity	SR2	—	—
Flammability classification	S3	—	—
Oxygen index	26.0%	—	—
Tensile strength (MPa)	33.7	2-6	≥16
Elongation at break (%)	385	>300	>400
Breaking force (N)	230	—	—
tear strength (N/m)	84	35	≥60
Acid resistance (5%)	No visible change	No visible change	No visible change
Alkali resistance (5%)	No significant change	No visible change	No significant change
Neutral salt spray (NSS)	No significant change	No visible change	No significant change
Water absorption (%)	≤5 (3.4%)	<5	≤5
Adhesion (MPa)	5	>2	≥2~2.5
Hardness (A)	15:90	Commonly	≥90
Sanding abrasion (L/μm)	≥1.42L/μm	No requirements	≤0.50cm^3/1.61km
VOC content (g/L)	158	<200	—
Proportion of solids (%)	>80	>80	>98
Liquid density (g/ml)	1.05	1.05	1.05~1.15
Viscosity (mPa·s)	14050	8000—20000	500~1500
Surface drying time (h)	≤2	6—12	≤120S
Actual drying time (h)	24	24	≤60s~10min

4. Construction technology

4.1. Surface treatment of lining concrete

(1) Manual cleaning

Loose animals, oxide skin, sand and floating dust on the surface can be cleaned manually with shovel, steel brush, sandpaper and other tools.

(2) Power tool cleaning

Power tools can be used to clean the surface deterioration layer and local concrete with low bearing capacity. The commonly used power tools include power wire brush, impact tool, grinder and grinder.

(3) Water washing and cleaning

Clean water is used to wash the soluble substances on the concrete surface. If there are salt and other soluble substances on the surface of the lining structure, protection should be taken to prevent the soluble substances from being transferred to the adjacent base surface along with the water flow, resulting in the occurrence of new soluble matter pollution.

4.2. Coating conditions and methods

(1) Application conditions

Painting work should only be carried out in good atmospheric environment and weather conditions with appropriate temperature. It is not suitable for coating under the following conditions: ① when the temperature drops to the lower limit of the drying or curing temperature of the coating; ② there is open water on the surface of the lining concrete; ③ when the humidity in the tunnel is greater than 85%.

(2) Painting method

Brush coating: paint brush made of high-quality natural fiber or artificial wool, with appropriate size and matching paint.

Roller coating: according to the type of coating and surface roughness, choose the roller with suitable pile length. Generally, phenolic roller core is used, and drum sleeve with short or medium long velvet is used. The drum sleeve should be cleaned before use, and loose fibers should be removed.

Scraping and coating: the scraper is used for manual coating, and the scraper is made of metal or non-metal.

Spraying: spray the coating by spray gun with low pressure air flow of compressed air.

According to the actual situation of the tunnel, it is suggested that brush coating combined with scraping method should be preferred in construction. Although this method is slow in efficiency, it can effectively reduce the workload of field facilities protection and avoid paint mist pollution caused by spraying.

4.3. Key points of construction

The construction shall meet the following requirements:

(1) The amount of curing agent should be prepared in strict accordance with the requirements of epoxy sealing, and sufficient curing time should be ensured after mixing evenly.

(2) During the construction of epoxy sealing primer, the first coat should be thinned and repeatedly applied to prevent missing coating, so as to ensure the penetration and infiltration effect of epoxy sealing primer on the concrete surface, so as to make it fully penetrate and seal the pores and cracks on the concrete surface.

(3) When applying one component modified polyurethane topcoat, it is necessary to ensure that the surface of epoxy sealing primer coating is dry. Before application, the water absorbent towel can be used to wipe and then apply, which can effectively prevent foaming between the two coatings.

(4) In order to ensure that the normal operation of the line will not be affected during the construction of operating tunnel, the modified polyurethane topcoat should be given with enough surface drying time.

Due to the different regions of the tunnel, the process requirements are not the same. The specific construction technology should be in line with the actual situation of the tunnel location. According to the local climate environment, the internal characteristics of the tunnel and the window duration, the construction scheme with strong pertinence should be formulated to ensure the normal operation of the tunnel line and the smooth construction.

5. Conclusions

(1) Railway tunnel lining crack disease accounts for no less than 30%, concrete pitting and construction joint laitance falling off occur from time to time. By applying non-destructive protective materials on the concrete cracks and the outer surface of spalling area, strengthening the local concrete, preventing further crack development and falling block is the actual demand of current engineering maintenance.

(2) The developed polyamide curing coating has high adhesion to concrete substrate and high interlayer adhesion with modified polyurethane. The tensile strength between them is more than 5MPa, with low volatility and good applicability.

(3) The developed one-component polyurethane protective material has the characteristics of high adhesion, good elongation, large tensile strength and rapid surface drying, and has good coating adaptability, which can ensure the coating quality and construction efficiency.

(4) The developed polyamide curing coating agent and one component polyurethane protective material can be applied by brush coating, rolling coating, scraping coating and other processes, which is convenient for construction and is conducive to promotion.

Acknowledgements

This study was financially supported by the Technology Research and Development Program of the China Railway Corporation (K2018G013), and the Major fund project of China Railway Research Institute Group Co., Ltd. (2019YJ027). The authors want to express sincere gratitude towards the anonymous reviewers for their through and deep remarks, which have greatly helped improve the paper technically as well as editorially.

References

[1] Tian Siming Gong Jiangfeng 2020 Statistics of railway tunnels in China as of end of 2019 *Tunnel Construction* **40(2)** 292

[2] Ma Weibin Chai Jinfei 2019 Development status of disease detection, monitoring, evaluation and treatment technology of railway tunnels in operation *Tunnel Construction* **39(10)**1553-1562

[3] Fan Yongjie Research on defect treatment technology of cutting lining by arch waterproof board of high-speed railway tunnel *Municipal·traffic·water resources engineering design* **2020(14)** 91-94

[4] Yang Wanjie Cause analysis and treatment technology of void behind secondary lining of railway tunnel *Traffic construction* **2020(20)** 275-276

[5] Zhao Tiancheng Discussion on remediation measures for lining defects of high-speed railway tunnel *Municipal·traffic·water resources engineering design* **2020(03)** 101-103

[6] Zhang Yong Zhang Bin1 Liu Zhi 2020 etc Application progress of polyurethane materials in railway construction *Polyurethane industry* **35(02)** 4-7

Investigation of carbonation resistance of recycled aggregate concrete

Zhiheng Deng[1], Bing Liu[1] and Ying Huang[1,*]

[1] College of civil engineering and architecture, Guangxi university, Guangxi 530004, China

[*] Email: gxu_hy@163.com

Abstract. This paper studied the carbonation resistance of recycled aggregate concrete (RAC) designed by two mix proportion design methods through accelerated carbonation test. The investigated variables include water-cement ratio, replacement ratio of recycled coarse aggregate (RCA), and carbonation time. Based on the test results, the effects of water-cement ratio, replacement ratio of RCA, mix proportion design method, and carbonation time on the carbonation resistance of RAC were analyzed and discussed. The results showed that the mix proportion design method has a significant effect on the carbonation resistance of RAC. The replacement ratio of RCA and the water-cement ratio alternately affected the carbonation resistance of RAC. The replacement ratio of RCA is the major factor affecting the carbonation resistance of RAC. In addition, a carbonation model for RAC considering the effect of the mix proportion design method was proposed. The comparison of the test values and calculated values indicates that the proposed model can properly predict the carbonation depth of RAC.

1. Introduction

Recently, with the rapid growth of the economy and population, more and more buildings and infrastructures are being built. Concrete, as the most widely used building material, is being used heavily, leading to significant consumption of the natural aggregates. Thereby lead to problems that natural aggregates shortage and environmental population. Thus, it is urgent to explore an effective measure to solve the problems. Meanwhile, numerous construction wastes are being generated in the demolition of old buildings [1,2]. The improper disposal of these wastes will cause a series of environmental problems [3]. Recycled aggregate concrete (RAC) is produced with recycled aggregates instead of natural aggregates. The recycled aggregates are made by construction wastes. So, using RAC instead of conventional ordinary concrete in construction can not only alleviate the shortage of natural aggregates but also solve the disposal problems of numerous construction wastes, which accords with the idea of sustainable development.

Previous studies [4,5] have proved that it is feasible to use of RCA in substitution to natural aggregate in production of concrete. Cabral et al. [6,7] observed that the tensile strength, cube compressive strength and elastic modulus of RAC are less than that of ordinary concrete, and the effect of recycled fine aggregate (RFA) on the mechanical properties of RAC is also less than that of RCA. However, Butler et al. [8] pointed out that the RAC splitting tensile strength is similar to ordinary concrete. Lima et al. [9] found that the compressive strength and tensile strength of RAC decrease with the increase of the replacement ratio of RCA. Deng et al. [10-13] found that the quality of RCA has a significant effect on the mechanical properties of RAC. Kou et al. [14] also observed that the mechanical properties of RAC made from high-quality RCA are not poorer than ordinary concrete. However, some studies [15-17] reported that the compressive strength of RAC would not be affected by the RCA.

The durability is an important property for concrete [18]. Previous studies [19-21] pointed out that as the replacement ratio of RCA increases, the deterioration ratio of mechanical properties of RAC is lower than that of durability properties. Xiao et al. [22] found that the microstructure of RAC is more complex than that of ordinary concrete; because of that, there are more interfacial transition zones (ITZ) in RAC, and the old mortar on RCA has many initial damage microcracks. Otsuki et al. [23] pointed out that due to the effect of the old ITZ and old mortar in RCA, the carbonation depth and the chloride ion permeability of RAC are slightly higher than that of ordinary concrete. Bravo et al. [24] also found that the durability of RAC is poorer than ordinary concrete. But the durability of RAC made from different RCA is significantly different. Faella et al. [25] pointed out that the high porosity of RCA is the major reason causing the poor durability of RAC. Lovato et al. [7, 18, 26-28] found that the carbonation depth of RAC is higher than that of ordinary concrete under the same condition, and the effect of RFA is greater than that of RCA. Silva et al. [18] conducted that the carbonation depth of RAC would be twice as high as that of ordinary concrete when the replacement ratio of RCA reaches to 100%. However, some studies [29-32] found that the carbonation depth of RAC is no worse than that of ordinary concrete.

Summarizing the literature mentioned above, it is found that the previous study results on the carbonation resistance of RAC are high discrete, which is the result of used different RCAs and different mix proportion design methods. Although the carbonation resistance of RAC has been studied by some scholars, studies about the effect of mix proportion design method on the carbonation resistance of RAC and prediction model for the carbonation depth of RAC considering the effect of mix proportion design method are rare. Therefore, the present study aims to investigate the carbonation resistance of RAC and establish the corresponding carbonation depth prediction model; the investigated variables include the RAC mix proportion design method, RCA replacement ratio, and water-cement ratio. For this goal, a total of 150 specimens were designed and tested. Based on the test results, the effect of the RAC mix proportion design method, RCA replacement ratio, and water-cement ratio on the carbonation resistance of RAC were analyzed and discussed, respectively. In addition, a RAC carbonation depth prediction model considering mix proportion design method, RCA replacement ratio, and water-cement ratio was established. The results of this study can expand the database of RAC carbonation depth and help to further understand the carbonation resistance of RAC.

2. Materials and testing procedure

2.1. *Raw materials*

Gumiao Brand P.O 42.5 ordinary Portland cement produced by Huahong Cement Co., Ltd. Guangxi, China was used in this test, and Table 1 listed its main physical properties tested according to GB175 [33]. The mixing water was tap water. The crushed stone from the Gravel Plant in Nanning was used as the natural coarse aggregate (NCA). The RCA used was made from the abandoned concrete collected from Nanning Express Ring Road. The abandoned concrete used in this study has been discarded by the road for a long time and attached with a lot of dust and asphalt. An appropriate amount of sample for this study was collected randomly and transported to the laboratory. After the process of crushing, washing, and sieving, the RCA used in this study was finally obtained and shown in Figure 1. The natural fine aggregate (NFA) used was Yongjiang river sand obtained from the Gravel Plant in Nanning. The physical properties of RCA, NCA, and NFA were tested following the code JGJ 52 [34] and shown in Table 2.

Figure 1. Recycled coarse aggregates.

Table 1. Main parameters of P.O.42.5.

Ignition loss of cement /%	Sieve residue /%	Normal consistency /%	Setting time /(h:min)		Flexural strength /MPa		Compressive strength /MPa	
			Initial	Final	3 d	28 d	3 d	28 d
2.23	0.4	26.8	2:25	3:36	5.42	8.83	24.6	46.8

Table 2. Physical properties of aggregate.

Type of aggregate	Bulk density /(Kg/m^3)	Maximum size /mm	Water absorption /%	Moisture content /%	Crushing index /%	Fineness modulus
NAC	2670	31.5	1.35	0.46	13	-
RCA	2545	31.5	4.32	0.68	33.52	-
NFA	2610	4.75	1.04	-	-	2.8

2.2. *Mix proportions*

For RAC, there are three commonly used mix proportion design methods. The first one is to design the RAC as same as ordinary concrete. The second one is to design the RAC using pre-absorbent aggregate according to the ordinary concrete design method. The third one is to design the RAC based on the free water-cement ratio. According to previous studies on the mix proportion design of RAC [35-38], it is found that the performance and mechanical properties of RAC designed with different mix proportion design methods are different. In this study, for a better discussion on the effect of both water-cement ratio and RCA replacement ratio on the carbonation resistance of RAC, we designed the RAC specimens with the first and the third mix proportion design methods. The mix proportions used are shown in Table 3, and the additional water used in the third mix proportion design method was calculated by Eq. (1).

$$AW = m_{RA}(W_{RA} - W_h) \quad (1)$$

Table 3. Mix proportions.

Number	Water/Cement	RCA ratio /%	Water /kg	Additional water /kg	Cement /kg	NFA /kg	NCA /kg	RCA /kg	Slump /mm
R4500		0	209	0	465	628	1161	0	90
R4503		30	209	0	465	628	813	348	90
R4505	0.45	50	209	0	465	628	580.5	580.5	85
R4507		70	251	0	558	628	348	813	85
R4510		100	251	0	558	628	0	1161	80

R5500		0	209	0	380	709	1156	0	95
R5503		30	209	0	380	709	809	347	95
R5505	0.55	50	209	0	380	709	578	578	95
R5507		70	251	0	456	709	347	809	90
R5510		100	251	0	456	709	0	1156	90
R6500		0	222	0	342	772	1158	0	100
R6503		30	222	0	342	772	811	347	100
R6505	0.65	50	266	0	410	772	579	579	100
R6507		70	266	0	410	772	347	811	100
R6510		100	266	0	410	772	0	1158	95
F4500		0	209	0	465	628	1161	0	95
F4503		30	221.7	12.7	465	564	813	348	95
F4505	0.45	50	230.1	21.1	465	564	599	599	85
F4507		70	238.6	29.6	465	564	356	842	85
F4510		100	251.3	42.3	465	564	0	1198	85
F6500		0	222	0	342	772	1158	0	100
F6503		30	234.6	12.6	342	667	811	347	100
F6505	0.65	50	243.1	21.1	342	667	579	579	95
F6507		70	251.5	29.5	342	667	347	811	95
F6510		100	264.2	42.2	342	667	0	1158	95

Number Description: 'R' represents the recycled aggregate concrete designed following ordinary concrete, and 'F' represents recycled aggregate concrete designed based on free water-cement ratio. And the first two figures represent its water-cement ratio, and the last two figures represent its RCA replacement ratio, for example, 'R4510' represents RAC designed following ordinary concrete, and its water-cement ratio and RCA replacement ratio are 0.45 and 100%, respectively.

2.3. Specimen production and testing methods

2.3.1. Compressive specimens Prepare the concrete according to Table 3 and put it into 150 mm × 150 mm × 150 mm plastic cube molds when the slump of the fresh concrete is between 80 mm and 100 mm. Next, put them on a vibrating table to be shaken and levelled. The vibratory compacted specimens were placed in a standard concrete curing room for 24 hours and then demoulded. The specimens were then placed back in the curing room. Measured the compressive strength of specimens after 28 days following the code GB/T 50081 [39].

2.3.2. Carbonation specimens Carbonized specimens were made with the same method as compressive specimens, whose size was 100 mm × 100 mm × 300 mm and cured for 24h before demoulded. After demoulding, all specimens were sealed with epoxy resin except for the two pressure surfaces leaving for carbonation. Next, all treated specimens were carbonized in a carbonation chamber at a temperature of 20 ± 0℃, a carbon dioxide concentration of 20 ± 3%, and a relative humidity of 70 ± 5%. The carbonation depth of all specimens was tested according to GB/T 50082 [40] at the ages of 3 d, 7 d, 14 d, 28 d, and 56 d, respectively.

3. Test results and discussion

3.1. Effect of water-cement ratio

3.1.1. RAC designed following ordinary concrete Figure 2 shows the relationship between the carbonation depth and water-cement ratio of RAC designed according to the mix proportion design meth of ordinary concrete. It can be seen that the carbonation depth of RAC increases with the increase of water-cement ratio, regardless of the replacement ratio of RCA. The reason is that the greater the water-cement ratio, the more the micropores and defects in RAC. However, under the different RCA replacement ratios, the level of the effect of the water-cement ratio on the carbonation depth of RAC is different. When the RCA replacement ratio is less than 50%, the increasing rate of carbonation depth slows or remains basically unchanged with the increase of water-cement ratio. While the RCA replacement ratio becomes greater than 50%, the increasing rate of carbonation depth increases with the increase of water-cement ratio, and this phenomenon becomes more noticeable when the RCA replacement ratio reaches 100%. This is because the RCA has both negative and positive effects on the carbonation resistance of RAC. On the one hand, the high water absorption of RCA makes the ITZ in RAC more compact, which increases the carbonation resistance of RAC. On the other hand, the initial damage cracks and high porosity of RCA weaken the carbonation resistance of RAC. When the RCA replacement ratio is low, the positive and negative effects of RCA are roughly counterbalanced. But, in the case of a higher RCA replacement ratio, the negative effects of RCA are greater than the positive ones. It can also be noted that as the water-cement ratio increases, both of the carbonization depth and its increasing rate of RAC is lower than that of ordinary concrete. Because of that, the effective water-cement ratio of RAC designed following ordinary concrete is lower than that of ordinary concrete under the same nominal water-cement ratio.

3.1.2. RAC designed based on free water-cement ratio Figure 3 shows the relationship between the carbonation depth and water-cement ratio of RAC designed based on free water-cement ratio. It can be seen that the carbonation depth of RAC also increases with the increase of water-cement ratio. But, compared with the RAC designed following the ordinary concrete, the carbonation depth of RAC designed based on free water-cement ratio is higher.

3.2. Effect of RCA replacement ratio

3.2.1. RAC designed following ordinary concrete Figure 4 shows that when the water-cement ratio is 0.45, the effect of the RCA replacement ratio on the carbonation depth of RAC is insignificant. When the water-cement ratio is 0.55, the carbonation depth of RAC is lower than that of ordinary concrete and decreases with the increase of RCA replacement ratio. When the water-cement ratio is 0.65, the carbonation depth of RAC is also lower than that of ordinary concrete, but first decreases and then increases with the increase of RCA replacement ratio. The reason is that when the water-cement ratio is low, the positive effects of RCA due to its high water absorption and the negative effects of RCA due to its initial damage defects cancel each other out. But when the water-cement ratio is high, the positive effects of RCA is greater than its negative effects. Thus, the carbonation depth of RAC shows a decreasing trend with the increase of the RCA replacement ratio. However, if the RCA replacement ratio is too high, its negative effects will outnumber its positive effects gradually, which causes an increase in carbonation depth.

3.2.2. RAC designed based on free water-cement ratio Figure 5 shows that regardless of what the water-cement ratio is, the carbonation depth of RAC increases with the increase of RCA replacement ratio, and its increasing rate is faster when the water-cement ratio is lower. The reason is that the positive effects of RCA due to its high water absorption were cancelled by the additional water. But the negative effects of RCA due to its initial damage defects still work.

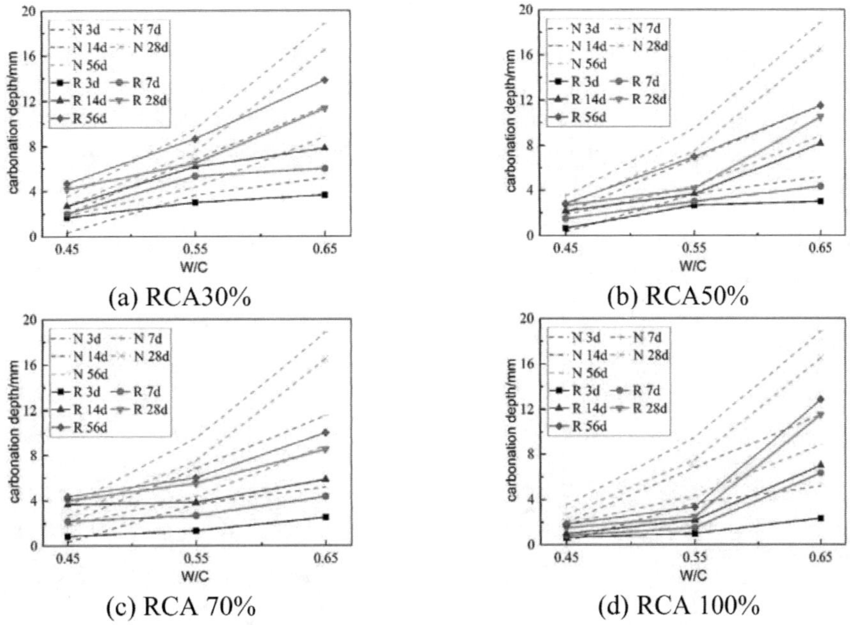

(a) RCA30% (b) RCA50%

(c) RCA 70% (d) RCA 100%

Figure 2. The relationship between the carbonation depth and water-cement ratio of RAC designed following ordinary concrete. (Note: In the legend, the prefix 'N' represents the ordinary concrete, i.e. RAC with an RCA replacement ratio of 0%; 'R' represents RAC).

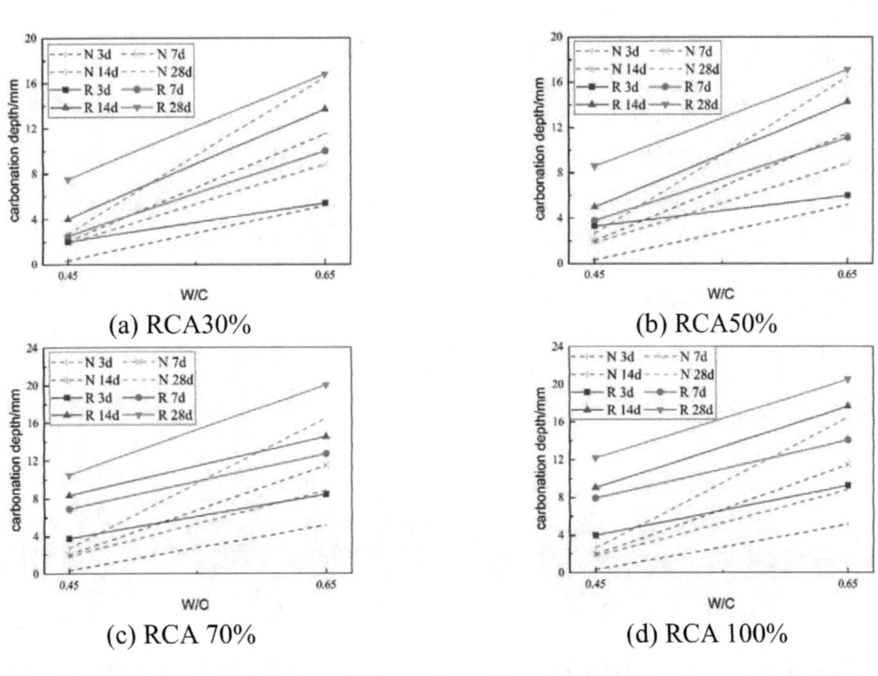

(a) RCA30% (b) RCA50%

(c) RCA 70% (d) RCA 100%

Figure 3. The relationship between the carbonation depth and water-cement ratio of RAC designed based on free water-cement ratio. (Note: In the legend, the prefix 'N' represents the ordinary concrete, i.e. RAC with an RCA replacement ratio of 0%; 'R' represents RAC).

IOP Conf. Series: Materials Science and Engineering 1028 (2021) 012014 doi:10.1088/1757-899X/1028/1/012014

3.3. Effect of mix proportion design method

Figure 6 shows the effect of the mix proportion design method on the carbonation depth of RAC. It can be seen that regardless of what the water-cement ratio is, the carbonation depth of RAC designed following ordinary concrete is much lower than that of RAC designed based on the free water-cement ratio. The higher the RCA replacement ratio, the greater the difference between the carbonation depth of the two types of RAC. The reason is that the effective water-cement ratio of the RAC designed following ordinary concrete is smaller than that of RAC designed based on the free water-cement ratio, and the higher the replacement ratio of RCA, the greater the difference between the effective water-cement ratio of the two types of RAC.

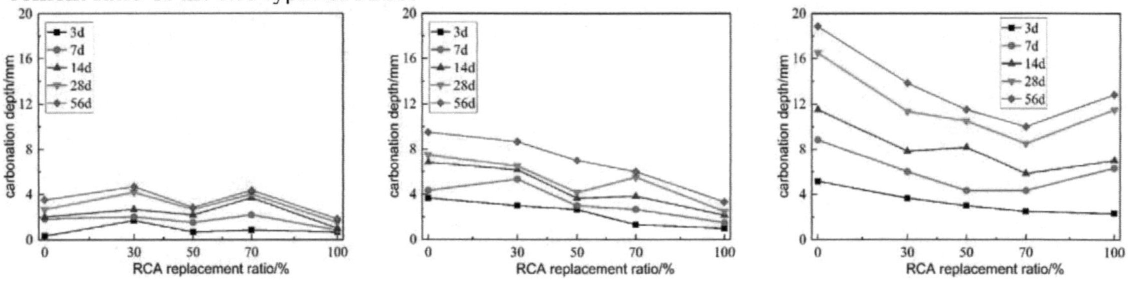

(a) W/C=0.45 (b) W/C=0.55 (c) W/C=0.65

Figure 4. The relationship between the carbonation depth and RCA replacement ratio of RAC designed following ordinary concrete.

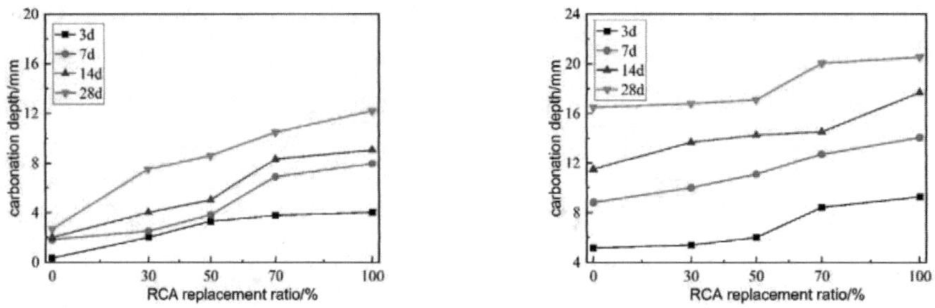

(a) W/C=0.45 (b) W/C=0.65

Figure 5. The relationship between the carbonation depth and RCA replacement ratio of RAC designed based on free water-cement ratio.

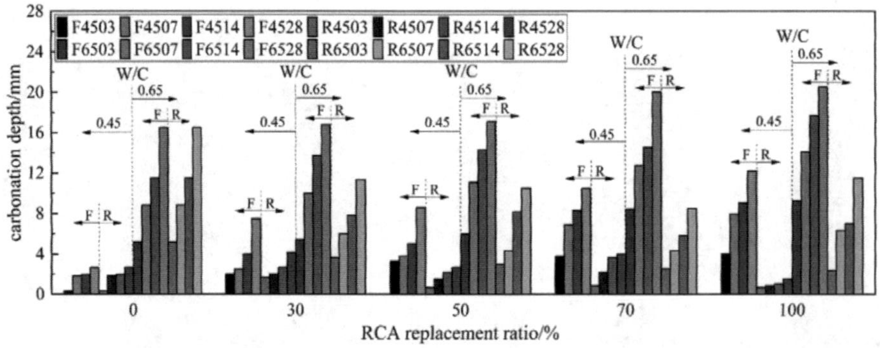

Figure 6. The comparison of the carbonation depth of RAC designed by the two mix proportion design methods. (Note: 'F' denotes the RAC designed based on free water-cement ratio; 'R' denotes the RAC designed following ordinary concrete, and the last two numbers denote the carbonation time).

3.4. Relationship between carbonation depth and time.

Figure 7 illustrates the relationship between the carbonation depth of RAC and carbonation time. It can be seen that the carbonation depth of RAC increases with the increase of carbonation time. Sagoe et al. [41] pointed out that the carbonation depth of RAC is a power function related to the carbonation time (Eq. (2)), and the majority of scholars believe that the exponent of the power function is 0.5. After fitting the test data using the power function Eq. (2), it is found that the power exponent is between 0.30 and 0.46, and the adjusted correlation coefficient R^2 is between 0.75 and 0.98, which indicates that Eq. (2) can well describe the relationship between the carbonation depth and carbonation time.

$$X_c = \alpha t^{\beta} \tag{1}$$

Where: α, β are undetermined coefficient.

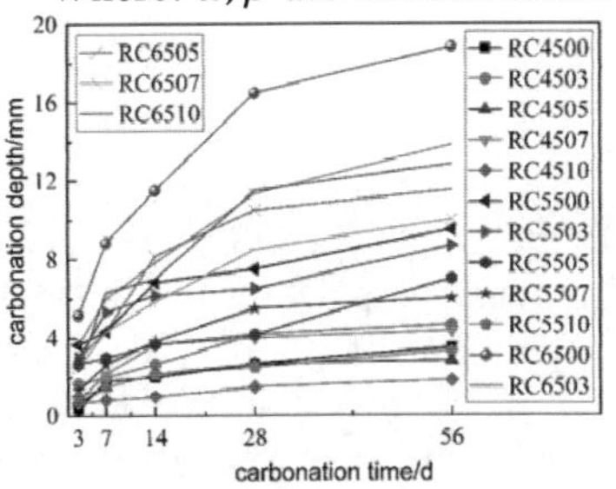

Figure 7. The relationship between the carbonation depth of RAC and carbonation time.

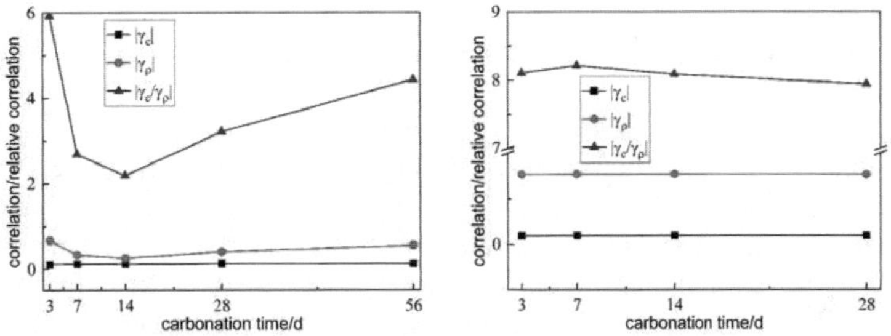

(a) RAC designed following ordinary concrete

(b) RAC designed based on free water-cement ratio

Figure 8. The relationship between the correlation (or relative correlation) and carbonation time.

3.5. Gray correlation analysis

As the carbonation process of RAC is affected by many factors and associated with uncertainty, it can be considered as a gray system [42], which means that the primary and secondary relationship of the effect of various factors on the RAC carbonation resistance can be analyzed by the gray correlation analysis. In this paper, an improved gray relational algorithm [43] was used to analyze the test data, and the calculation results are listed in Table 4. As shown in Table 5, when RAC designed following ordinary concrete, the correlation of water-cement ratio and RCA replacement ratio to the carbonation depth of all ages specimens are positive and negative, respectively. It indicates that the water-cement ratio is positively correlated with the carbonation depth of RAC, while the RCA replacement ratio is negatively correlated with the carbonation depth of RAC. When RAC designed based on the free water-cement ratio, both of the correlation of water-cement ratio and RCA replacement ratio to the carbonation depth of all ages specimens are positive. It indicates that both of the RCA replacement ratio and water-cement ratio are positively correlated with the carbonation depth of RAC. These conclusions are consistent with these conclusions of the qualitative analytical conclusions obtained above. The value of $|\gamma_p/\gamma_C|$ is all greater than 1, which indicates that for the two mix proportion design methods of RAC, the replacement ratio of RCA is the primary factor that influences the RAC carbonation depth, and the water-cement ratio is the subordinate factor. Figure 8 shows the relationship between the correlation (or relative correlation) and carbonation time. It can be seen that for the two mix proportion design methods of RAC, the correlation of water-cement ratio to RAC carbonation depth changes very little with the increase of the carbonation time, which means that the effects of water-cement ratio on the carbonation depth of RAC in different carbonation times are basically the same. When RAC designed following ordinary concrete, the relative influence of water-cement ratio and RCA replacement ratio on the carbonation depth of RAC first decreases and then increases with the increase of the carbonation time. However, when RAC designed based on the free water-cement ratio, the opposite trend is shown.

Table 4. Correlation.

Age/d	RAC designed following ordinary concrete			RAC designed based on free water-cement ratio		
	γ_C	γ_P	$\lvert\gamma_P/\gamma_C\rvert$	γ_C	γ_P	$\lvert\gamma_P/\gamma_C\rvert$
3	0.114	-0.674	5.926	0.095	0.767	8.107
7	0.126	-0.339	2.691	0.094	0.770	8.216
14	0.119	-0.261	2.190	0.095	0.769	8.085
28	0.128	-0.410	3.213	0.096	0.766	7.941
56	0.125	-0.554	4.429	-	-	-

4. Recycled aggregate concrete carbonation model

4.1. Establishment of the carbonation model

Chinese Standard "Standard for durability assessment of concrete structures" [44] gives the ordinary concrete carbonation model. However, the microstructure of RAC is more complicated than that of ordinary concrete, and the influential factors of RAC are also more than that of ordinary concrete. Thus, the carbonation model of ordinary concrete is not suitable for RAC. According to the previous analysis, the carbonation depth of RAC is a power function relationship with the carbonation time. The results of this study and some other studies [45-47] also indicate that all of the RCA replacement ratio, water-cement ratio, mix design method are significantly affect the RAC carbonation resistance. Thus, taking these factors into account, a correcting model Eq. (3) based on this standard is obtained.

$$X_c = K_{RA}K_{co_2}K_{k1}K_{ks}T^{0.25}R^{1.5}(1-R)(\frac{A}{K_p f_{cu}^{RC}} + B)t^C \tag{3}$$

Where:$K_{RA} = \alpha\gamma + \beta$; $K_{co_2} = \sqrt{\frac{C_0}{0.2}}$; corner area take K_{kl}=1.4, non-corner area take K_{kl}=1.0; under pressure take K_{ks}=1.0, under tension take K_{ks}=1.7; A、B、C are undetermined coefficient.

According to the previous analysis, the rule of carbonation resistance of RAC with the RCA replacement ratio of less than 50% and more than 50% are different. So, the authors use Eq. (2) to fit the test data with a replacement ratio of RCA subdivided by 50%. The undetermined parameters and the recommended K_p values in the RAC carbonation depth prediction model are shown in Table 5. As shown in Table 6, the predictive values agree well with the test value. It indicates that the established carbonation model can effectively model the carbonation resistance of RAC.

Table 5. Model parameters and the K_p recommended values.

Replacement ratio	α	β	A	B	C	K_p		
						P1	P2	P3
Not more than 50%	-0.517	102.939	4.989	-0.144	0.383	1.000	0.899	0.571
Greater than 50%	0.794	35.712	5.294	-0.136	0.426	1.000	0.920	0.517

Table 6. Comparison of the predictive values and the test values.

Number	7d			28d		
	Test value/E	Model value/LM	LM/E	Test value/E	Model value/LM	LM/E
R4500	1.83	1.59	0.87	2.67	2.71	1.02
R4503	2	1.96	0.98	4.17	3.34	0.80
R4505	1.5	0.76	0.51	2.67	1.30	0.49

R4507	2.17	0.92	0.42	4.00	1.65	0.41
R4510	0.83	1.29	1.55	1.50	2.32	1.55
R5500	4.33	5.27	1.22	7.50	8.97	1.20
R5503	5.33	4.75	0.89	6.50	8.08	1.24
R5505	3	3.49	1.16	4.17	5.94	1.43
R5507	2.67	2.55	0.96	5.50	4.61	0.84
R5510	1.5	2.05	1.37	2.50	3.71	1.48
R6500	8.83	8.38	0.95	16.50	14.25	0.86
R6503	6	5.67	0.95	11.33	9.65	0.85
R6505	4.33	5.29	1.22	10.50	8.99	0.86
R6507	4.33	4.57	1.06	8.50	8.26	0.97
R6510	6.33	5.16	0.82	11.50	9.32	0.81
F4500	1.83	3.08	1.68	2.67	5.23	1.96
F4503	2.5	2.95	1.18	7.50	5.02	0.67
F4505	3.8	4.03	1.06	8.60	6.86	0.80
F4507	6.88	5.17	0.75	10.48	9.34	0.89
F4510	7.95	7.14	0.90	12.20	12.88	1.06
F6500	8.83	10.62	1.20	16.50	18.07	1.10
F6503	10	9.10	0.91	16.80	15.47	0.92
F6505	11.1	8.65	0.78	17.11	14.72	0.86
F6507	12.72	10.32	0.81	20.04	18.62	0.93
F6510	14.08	13.50	0.96	20.54	24.37	1.19
E1	10.1	13.73	1.36	16.50	23.34	1.41
E2	9.5	7.73	0.81	17.00	13.15	0.77
E3	12.2	7.69	0.63	22.00	13.07	0.59
E4	16.2	8.95	0.55	26.00	16.15	0.62
E5	10.4	12.27	1.18	17.10	22.14	1.29
D1	10.4	12.27	1.18	17.10	22.14	1.29
D2	14.7	15.93	1.08	23.10	28.75	1.24
D3	7.3	12.06	1.65	29.20	21.77	0.75
D4	6.3	11.92	1.89	25.80	21.52	0.83
Average			1.04			1.00
Standard deviation			0.33			0.33
Coefficient of variation			0.32			0.33

4.2. *Verification of the carbonation model*

To verify the carbonation model established above, the experimental data of literature [46] and [47] were selected. After calculating, it is found that the adjusted correlation coefficient between the predictive values and test values is R^2=0.959, which verifies the rationality and effectiveness of the carbonation model established above for RAC.

5. Conclusions

This study investigated the carbonation resistance of RAC designed by two mix proportion design by accelerated carbonation test. The effects of carbonation time, mix proportion design method, RCA replacement ratio, and water-cement ratio on the carbonation resistance of RAC were analyzed and discussed. Based on the test results and discussions, the conclusions can be drawn as follows:

The carbonation depth of RAC increases with the increase of water-cement ratio, and the carbonation depth of RAC with a higher water-cement ratio is more sensitive to the RCA replacement ratio than that of RAC with a lower water-cement ratio.

The effect of RCA replacement ratio on the RAC carbonation resistance is not always monotonous, which is influenced by the water-cement ratio and mix design method. When the RAC is designed following ordinary concrete and the water-cement ratio is lower, the RAC carbonation depth varies little with the RCA replacement ratio, but when the water-cement ratio is higher, the RAC carbonation depth decreases with the increase of the RCA replacement ratio. When the RAC is designed based on the free water-cement ratio, the carbonation depth of RAC increases with the increase of the RCA replacement ratio.

The mix proportion design method of RAC affected its carbonation resistance. The carbonation resistance of RAC designed following ordinary concrete is better than that of RAC designed based on the free water-cement ratio.

The RCA replacement ratio is the primary factor that influences the carbonation resistance of RAC regardless of its mix proportion design method.

The carbonation depth of RAC has a power function relationship with the carbonation time, and its power exponent is between 0.30 and 0.46.

A carbonation model for RAC considering the effect of the RCA replacement ratio and mix proportion design method was established. The established model can properly predict the carbonation depth of RAC.

Acknowledgments

The authors gratefully acknowledge the financial supported by the National Natural Science Foundation of China (Grant No.51478126 and Grant No. 51868005).

6. Appendix A

List of notations

AW is the additional water (Kg)	t is the carbonation time
m_{RA} is the mass of RCA (Kg)	γ is replacement ratio of RCA
W_h is the water content of RCA (%)	C_0 is the volume of carbon dioxide
W_{RA} is the water absorption of RCA (%)	R is relative humidity
Xc is the carbonation depth	T is environment temperature
K_{RA} is the influence coefficien of RCA replacement ratio	K_p is the influence coefficient of mix design method
K_{co_2} is the influence coefficient of carbon dioxide concentration	${f_{cu}}^{RC}$ is the average compressive strength of RAC
K_{kl} is the influence coefficient of position	

References

[1] S. Kou, C. Poon, H. Wan 2012 *Constr build mater* **36**

[2] V. W. Y. Tam, C. M. Tam, Y. Wang 2007 *Constr build mater* **21** 10
[3] M. Behera, S. K. Bhattacharyya, A. K. Minocha, et al. 2014 *Constr build mater* **68**
[4] H. Dilbas, M. şimşek, Ö. çakır 2014 *Constr build mater* **61**
[5] M. Bravo, J. D. Brito, J. Pontes, et al. 2015 *Constr. Build. Mater* **77**
[6] A. E. B. Cabral, V. Schalch, D. C. C. D. Molin, et al. 2010 *Constr. Build. Mater* **24** 4
[7] P. S. Lovato, E. Possan, D. C. Coitinho Dal Molin, et al. 2012 *Constr. Build. Mater* **26** 1
[8] L. Butler, J. S. West, S. L. Tighe 2013 *Constr. Build. Mater* **47**
[9] C. Lima, A. Caggiano, C. Faella, et al. 2013 *Constr. Build. Mater* **47**
[10] Z. Deng, B. Liu, B. Ye, et al. 2019 *J. Mater. Cycles Waste* 22
[11] B. Liu, C. Feng, Z. Deng 2019 *Constr. Build. Mater* **217**
[12] W. H. Kwan, M. Ramli, K. J. Kam, et al. 2012 *Constr. Build. Mater* **26** 1
[13] A. K. Padmini, K. Ramamurthy, M. S. Mathews 2009 *Constr. Build. Mater***23** 2
[14] S. Kou, C. Poon 2015 *Constr. Build. Mater* **77**
[15] N. Fonseca, J. D. Brito, L. Evangelista 2011 *Cement Concrete Comp***33** 6
[16] A. A. Elhakam, A. E. Mohamed, E. Awad 2012 *Constr. Build. Mater***35** none
[17] S. B. Huda, M. S. Alam 2014 *Constr. Build. Mater* **65**
[18] R. V. Silva, R. Neves, J. de Brito, et al. 2015 *Cement Concrete Comp* **62**
[19] S. W. Tabsh, A. S. Abdelfatah 2009 *Constr. Build. Mater* **23** 2
[20] C. Thomas, J. Setién, J. A. Polanco, et al. 2013 *Constr. Build. Mater* **40**
[21] A. Kanellopoulos, D. Nicolaides, M. F. Petrou 2014 *Constr. Build. Mater* **53** 4
[22] J. Xiao, W. Li, Y. Fan, et al. 2012 *Constr. Build. Mater* **31**
[23] N. Otsuki, S. Miyazato, W. Yodsudjai 2003 *J. Mater. Civil Eng* **15** 5
[24] M. Bravo, J. de Brito, J. Pontes, et al. 2015 *Constr. Build. Mater* **77**
[25] C. Faella, C. Lima, E. Martinelli, et al. 2016 *Cement Concrete Comp* **71**
[26] V. Carević, I. Ignjatović, J. Dragaš 2019 *Constr. Build. Mater* **213**
[27] R. Kurda, J. de Brito, J. D. Silvestre 2019 *Journal of CO2 Utilization* **29**
[28] S. R. D. Silva, J. J. D. O. Andrade 2017 *Constr. Build. Mater* **153**
[29] S. M. Levy, P. Helene 2004 *Cement & Concrete Research* **34** 11
[30] G. Fathifazl, A. Abbas, A. G. Razaqpur, et al. 2009 *J. Mater. Civil Eng* **21** 10
[31] S. Levy, P. Helene 2007 *Exacta* **5** 1
[32] R. Mi, G. Pan 2020 *Journal of Harbin Engineering University*
[33] GB 175-2007 2007 Beijing China
[34] JGJ 52-2006 2006 Beijing China
[35] J. Shen, Y. Xu, S. Zhou, et al. 2007 *New building materials* 08
[36] N. K. Bairagi, H. S. Vidyadhara, K. Ravande 1990 *Constr. Build. Mater***4** 4
[37] S. Song, L. Wang 2009 *Journal of Wuhan University of Technology***31** 07
[38] W. Bai, J. Zhang 2007 *Journal of Shaoxing University* 03
[39] GB/T50081-2002 2002 Beijing China
[40] GB/T50082-2009 2002 Beijing China
[41] K. K. Sagoe-Crentsil, T. Brown, A. H. Taylor 2001 *Cement Concrete Res* **31** 5
[42] X. Tan, J. Deng 1995 *Statistical Research* 03
[43] H. Yuan, M. Yang, F. Li, et al. 2005 *Journal of Wuhan University of Technology* 02
[44] CECS220-2007 2007 Beijing
[45] O. Geng, X. Zhang, C. Zhang 2015 *Journal of China University of Mining and Technology* **44** 01
[46] W. Xiao, Z. Guo, Z. Wu, et al. 2015 *Journal of Architectural and Environmental Engineering* 06
[47] B. Lei, J. Xiao 2008 *Journal of Building Materials* 05

IOP Conf. Series: Materials Science and Engineering 1028 (2021) 012015 doi:10.1088/1757-899X/1028/1/012015

Effect of Ni on the oxygen reduction reaction for low alloy steels in thin electrolyte layer conditions

Shan Jiang*, NingCai, Jianping Cao and Jianwei Yang

Shougang research institute of technology, ShougangGroupCo.,Ltd., China.

*Email: jiangshan@shougang.com.cn

Abstract. The corrosion behavior of carbon and a Ni-containing steel is investigated in the environment of a thin electrolyte layer. Potentiodynamiccathodic polarization curves were determined using a thin electrolyte layer apparatus at ordinary and negative pressure environments. Oxygen reduction is an important cathodic reaction for carbon steel in the present study. However, the suppression of oxygen reduction occurred for Ni-containing steels. The inhibition of oxygen reduction was closely correlated with the formation of the amorphization of the corrosion product, which contained amount of Ni, formed on the Ni-containing steel.

1. Introduction

Low alloy steels are widely used in many application fields, such as construction, bridges and pipelines. The corrosion resistance of low alloy steels has attracted significantattention. Weathering steel is a kind of low alloy steel thatisappliedin construction and bridges. The nature of the corrosion product formed on weathering steel are receiving increasing attention [1–7]. Compared with carbon steel, the enhanced corrosion resistance of weathering steel is due to the compact rust layer formed on the surface. However, the corrosion of steel in marine atmosphere exposure is much higher than steel in other environment exposure.In recent years, anewtype of advanced weathering steel containing 3 wt.% Nihas been developed and applied to bridges in some coastal areas where the air born salinity is higher than the upper limit for weathering steel [8].

Kimura[9] suggested that the excellent corrosion resistance of advanced weathering steelwhich contained 3wt%Ni, isachieved by adjustment the $Fe(O,OH)_6$lattic structures of rust formed onits surface. The stabilized Fe_3O_4,i.e., Fe_2NiO_4, in the nano-network changesthe ion-exchanging properties of rustfrom anion to cationselective.Similar results have also been reported by other researchers [10].

Chenet al.[11] suggested that Ni is enrichedin the rust layer formed on weathering steel, especially in the inner rust layer, while Cl^- almost concentrated in the outer rust layer. Therefore, it was speculated that the cationselective property of the rust layer of Ni-containing weathering steelcan be achieved by Ni concentration(the content of Ni exceeds 4 wt.%).

Diaz et al. [12] suggested that the higher content of Ni leads to higher corrosion resistance of advanced weathering steel. It can be attributed to the high proportion of nano-phasic goethite (particle size smaller than 15 nm) in the inner rust layer by concentration of Ni in the rust layer.These results are consistent with findings of Cano et al. [13],who pointed out that high Ni content steels exhibited excellent corrosion resistance after a 2-year.

It is well-known that the corrosion resistance of lowalloysteelscan be improved by increasing their nickel content[12,14,15]. However, according to the research of Cheng[16], when the Ni content in weathering steels exceeds 3.5 wt.%, the corrosion resistance improves slightly even with a considerable increase of Ni content.Therefore, 3.5 wt.% is the optimal Ni content in weathering steels to decrease the corrosion rate inindustrail atmosphere.

In most previous studies, the ionselective property of rust layers containing Niand the effect of Ni content on the corrosion resistance of low alloy steel have been the focus, with reports describing the influenceof Ni on the electrochemical reactions of low alloy steels and the characteristics of corrosion products being rare.In the present study, the thin electrolyte layer method is preceded on a solution containing Cl^{-}. The effects of Ni on theoxygen reduction reaction and on the characteristics of rust layer formation on low alloy steels are also analyzed. The propose of this study was to understand the role of Ni in improving the corrosion resistance of low alloy steel exposed in marine atmospheres with high humidity.

2. Experimental

2.1. Thin electrolyte layer apparatus

A thin electrolyte layer apparatus has been utilizedin previous studies and the theory of the measurement of the electrolyte layer thickness that established a loop connected by a Pt needle, working electrode and ampere meter has been discussed[17–19].According to this method, the distance between the Pt needle and electrode surface wasdetermined to be 100μmin the present study. As shown in Figure 1, the apparatusdiscussed in previous studies was put into a chamber thatisolated the air.An air pump was connected to the chamber. Therefore, the apparatus can prevent a negative pressure environment for the experiment.

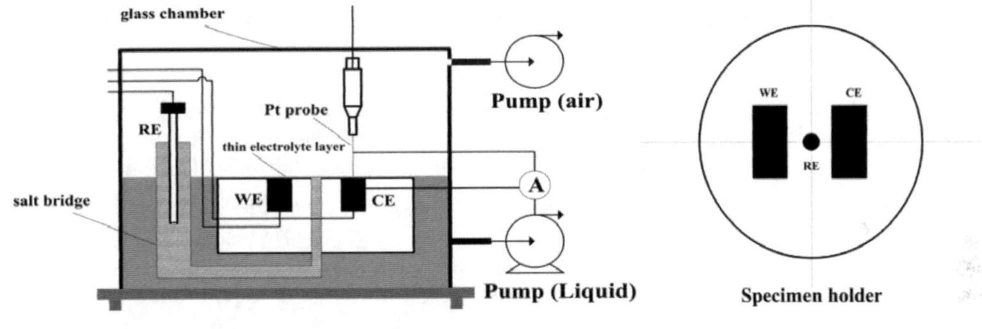

Figure 1. Schematic of thin electrolyte layer apparatus.

The thickness of the thin electrolyte layer decreased with the time increasing. Once the thickness of the thin electrolyte layer was too low to touch the Pt needle, the current could not be observed by an ohmmeter. At this time, the peristaltic pump was turned on to put into the water until the current was observed by the ohmmeter alarm.It must be noted thatwhen the thickness of the thin electrolyte layer was less than 90μm, the current was not observed because of the water tension. Therefore, the thickness of the thin electrolyte layer was controlled at the range of 85–110μm.

The experiments were performed in a NaCl solution (pH6.8) at a temperature of 25°C and the concentration of NaCl was 0.62mol/L(mass percent of 3.5 wt.%). The thickness of the thin electrolyte layer on the electrode surface approached 100μm.

2.2. Specimens

Both of the Ni-containing steel and carbon steel were corroded in the thin electrolyte layer environment. Table 1 shows the chemical composition of experimental steels. The Dimension of exposure area was 12mm×8mm.

Table 1. Chemical composition of experimental steels, wt%.

	C	Si	Mn	P	S	Ni	Fe
Carbon steel	0.060	0.26	1.01	<0.01	≤0.005	-	Bal.
Ni-containing steel	0.054	0.31	0.98	<0.01	≤0.005	1.8-2.1	Bal.

2.3. Weight loss
Before corrosion test, the specimens were polished by emery paper, and then rinsed with ethyl alcohol. Samples after cleaning were weighed with a electronic microbalance. After corrosion test, the specimens were acid cleaning by immersed in pickling solution(500 mL HCl + 3.5g hexamethylenetetramine) and then specimens were cleaned by distill water and ethyl alcohol. Finally, the specimens were weighed again so that the weight loss of each specimens were calculated leading by corrosion test.

2.4. Electrochemical measurements
Potentiodynamic polarization test were carry out by using a Princeton Applied Research Model 4000+ potentiostat, and the scanning rate was determined as 1mV/s. Saturated calomel electrode was used as reference electrode.

2.5. Rust layer analysis
The corrosion products formed on Ni containing steel and carbon steel were removed from the steel surface, and then mechanical ground into a powder. The corrosion product powdered samples were used for XRD analyses. The XRD(Bruker Model D8 Advance, Cu target) measurements were carried out by a scanning rate at $1° \min^{-1}$.

The distribution of Ni in the rust layer was determined by using a cross-section specimen. The specimens containing rust layer were fixed by an epoxy resin. The cross-section specimens of rust layer of the fixed specimens were were polished by emery paper grades form150# to 1000#. Then, the cross-section of the corrosion products was characterized via electron probe micro-analysis(EPMA),Shimadzu Model EPM810 electron probe micro-analyzer (accelerating voltage: 15 kV, sample current: 0.1 μA).

The corrosion product from inner rust layer was separated from steel surface via ultrasonic vibration in ethyl alcohol for 1h. The ethyl alcohol containing corrosion product was dripped on a carbon film. After the ethanol evaporated, corrosion products were attached to carbon film. TEM analysis of the film were carried out on a FEI FECNAI G2 F20 microscope.

3. Results and discussion

3.1. Weight loss of experimental steels
Figure 2 shows the results of weight loss of experimental steels. In the initial stage ,there was no significant difference in weight loss between the carbon steel and Ni containing steel. After 48h of corrosion, the carbon steel exhibited the highest weight loss. Theweight loss of Ni-containing steel was accordingly less than that of the carbon steel, and the difference of weight loss between carbon steel and Ni containing steel increases with increasing corrosion period.

3.2. Potentiodynamic polarization curves
Figure 3 shows the cathodicpotentiodynamic polarization curves of experimental steels tested in an ordinary pressure environment (with O_2) and in a negative pressure environment(without O_2). The current density associated with the cathodic shift of carbon steel tested in the ordinary pressure environment was much higher than that in the negative pressure environment. This phenomenon suggested that oxygen reduction played an important role in the cathodic reaction occurring on the rust layer formed on the carbon steel. However, the same phenomenon was not observed for Ni-containing

steel. There was almost no difference in current density between the Ni-containing steel tested in the two environments.

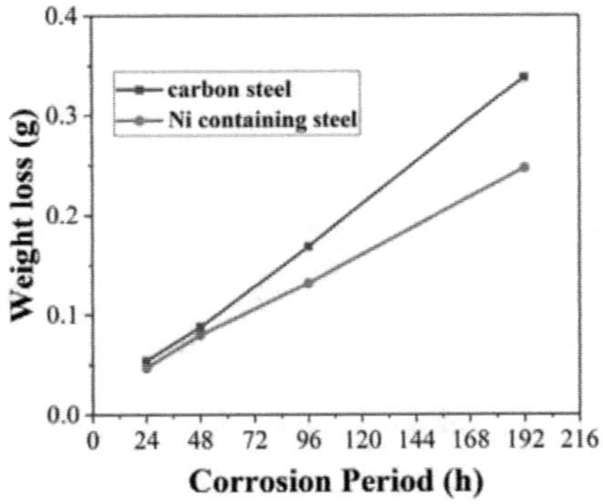

Figure 2. Weight loss of experimental steels.

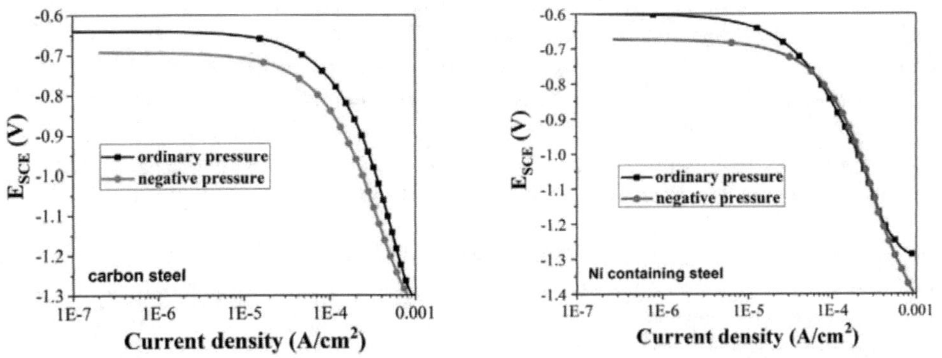

Figure 3. Cathodicpotentiodynamic polarization curves of experimental steels tested in an ordinary pressure environment and in a negative pressure environment.

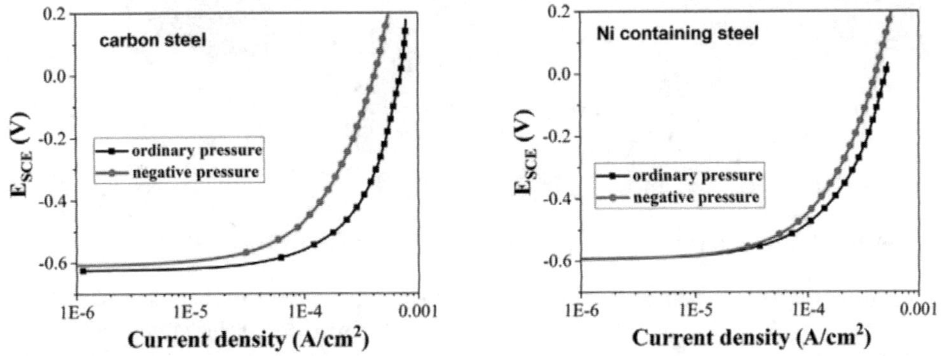

Figure 4. Anodicpotentiodynamic polarization curves of experimental steels tested in an ordinary pressure environment and in a negative pressure environment.

Figure 4 shows the anodicpotentiodynamic polarization curves of experimental steels tested in the two environments (with and without O_2).Observably, the anodic current density of carbon steel tested in the ordinary pressure environment was higher than that in the negative pressure environment. While,

IOP Conf. Series: Materials Science and Engineering 1028 (2021) 012015 doi:10.1088/1757-899X/1028/1/012015

the anodic current density tested in the ordinary pressure environment, for Ni-containing steel, was also higher than that in the negative pressure environment, but the difference between the current densities tested in the two environments was not obvious. These results indicate that oxidation of corrosion product formed on Ni containing steel was inhibited in the ordinary pressure environment.

It was clear that the main difference of electrochemical behavior between carbon and Ni-containing steel can be attributed to the addition of Ni in the Ni containing steel.

3.3. Distribution of Ni in the rust layer of the Ni-containing steel

Figure 5 shows the distribution of Ni in the rust layer formed on the Ni-containing steel. And these results suggest that Ni is enriched in the rust layer, and Ni content of the inner rust layer is a little higher than those of steel substrate and the outer rust layer.Therefore, the rust layer formed on Ni-containing steel existed concentration behavior of Ni[9,11–13]. The difference of electrochemical behavior between carbon steel and Ni containing steel was attributed to the amount of Ni in the rust layers.

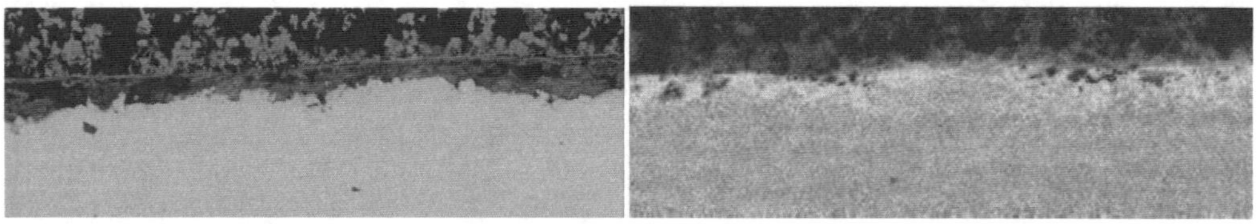

Figure 5. Distribution of Ni in the rust layer formed on Ni-containing steel.

3.4. TEM analysis of rust layer formed on Ni-containing steel

A TEM image and EDS analysis of the inner rust layer corrosion product formed on Ni-containing steel is shown in Figure 6. The corrosion product revealed two remarkably different areas in terms of both composition and crystallinity, i.e.,aNi-containing area and anon-Ni-containing area. In the case of the non-Ni-containing area, the product gave rise to shape diffraction patterns. In contrast, in the case of the Ni-containing area, the corrosion product gave rise to a diffraction pattern of diffusion ring,which may indicate the existing of amorphous substance. Therefore, it can be speculated that Ni enrichment of the rust layer leads to the amorphization of the corrosion products.

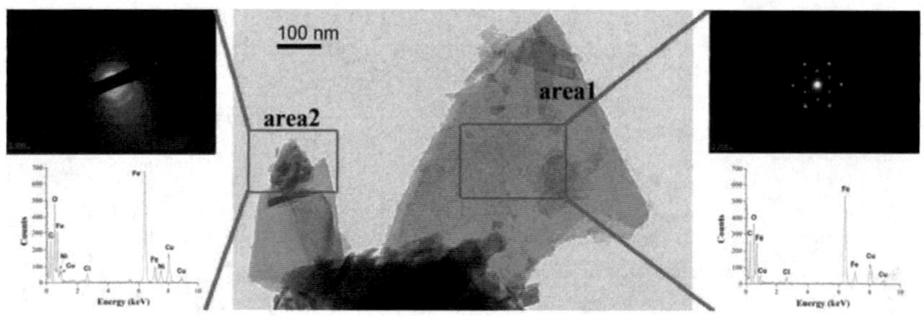

Figure 6. TEM analysis of corrosion products formed on the Ni containing steel.

3.5. XRD analysis of rust layer formed on experimental steels

Figure 7 shows the results of XRD analysis of corrosion products formed experimental steels. Both of the corrosion products formed on carbon steel and Ni containing steel were consist of α-FeOOH, γ-FeOOH and Fe_3O_4, and Fe_3O_4 is the main corrosion product in the rust layers by comparing the intensity of the diffraction peaks. The main difference of the two corrosion products is that the Fe_3O_4 diffraction peaks of Ni containing steel were moving to a few higher diffraction angles than that of carbon steel. It is indicated that formation of $NiFe_2O_4$ may occur during corrosion process, in other words, Ni may distributed in the Fe_3O_4 in the rust formed on Ni containing steel. Therefore, the amorphous substance may consist of Fe_3O_4 doped Ni.

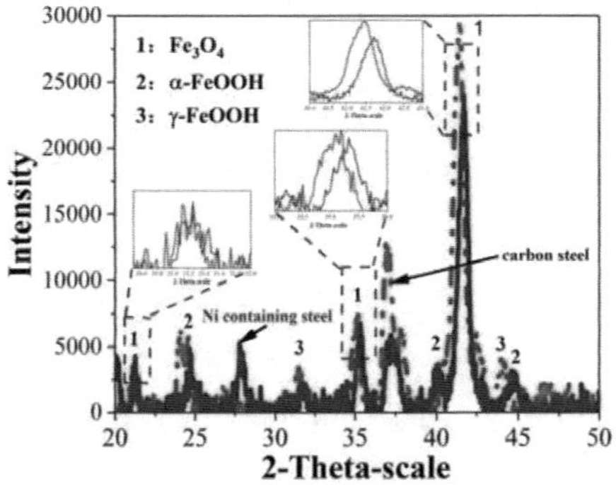

Figure 7.XRD analysis of corrosion products formed on the experimental steels.

4. Discussion

The main difference in corrosion behavior between carbon and Ni-containing steels can be attributed to the corrosion product formed on the steels. As is mentioned in the previous study[20], magnetite and maghemitecoexisted as a mixture of phases in the rust layer in desaeratedcorrosion medium, while lepidocrocite(γ-FeOOH) or ferrihydrite formation can be occured in aerated medium. Therefore, the difference of oxidation behavior of corrosion product formed on Ni containing steel between negative pressure environment and ordinary pressure environment was lower than that of carbon steel. It is indicated that formation of lepidocrocite or ferrihydrite in rust layer formed on Ni containing steel is more difficult than that in rust layer formed on carbon steel. However, this anodic electrochemical behavior may not impact corrosion rate of the steels directly.

Therefore, the rust layer plays an important role in O_2reduction which influence the corrosion rate of the steels directly. The rust layer is a type of porosint. There are many pores distributed on the surface of the rust layer. According to Hoerlè[21] and Stratmann[22,23], when a certain amount of the γ-FeOOH is reduced to the conducting substance, namely reduced lepidocrocite({FeOHOH}).The crystallographic structureof reduced lepidocrocite is consist withthat of γ-FeOOH[24]. It coats the surface developed by the rust layer porosity. If a connection between the underlying metal exists, then the {FeOHOH} on the surface of the rust can act as a cathodic area. Oxygen reduction occurred on the surface of rust layer and on the inwall of the pores.

The thin electrolyte layer condition is similar with the condition of beginning of the drying stage of wet-dry cycles. Therefore, the oxygen diffusion can maintain a high rate through the corrosion period, leading to the high rate of oxygen reduction on surface of rust containing {FeOHOH}. The corrosion product acted as a wire to transport electrons from the metal surface to the surface of rust layer and the inwall of the pores. The corrosion product formed on the Ni-containing steels contained a certain amount of $NiFe_2O_4$, which belongs to the magnetite. As previous studymentioned[12], some nanosized corrosion products with a particle size smaller than 15 nm were attributed to the concentration of Ni in the inner rust layer. Owing to the formation of nanosized Ni-containing corrosion products, the incorporation of Ni into the Fe_3O_4 may lead to the formation of amorphizationofthe corrosion product, thereby decreasing the electronic conductivity of the corrosion product. Because of the poor conductivity of the amorphous corrosion product containing Ni, the process of transporting electrons was inhibited, thereby inhibiting the oxygen reduction reaction.

For the thin electrolyte layer conditions, the oxygen diffusion can maintain a high rate throughout the corrosion period. The concentration of Ni in the rust layer led to the inhibition of oxygen reduction occurring on the surface of the rust layer. Therefore, the addition of Ni can improve the corrosion resistance of low alloy steel exposed to thin electrolyte layer environments.

Some aspects need to be understood clearly. For example, the proportion of oxygen reduction and corrosion product reduction occurring on rust formed on the carbon and Ni-containing steels, and the effect of Ni on the oxidation reaction of rust product should also be investigated in further detail.

5. Conclusions

The corrosion behavior of carbon and Ni-containing steel was investigated during thin electrolyte layer conditions with and without O_2. The main findings may be summarized as follows:

Theweight loss of Ni-containing steel was accordingly less than that of the carbon steel in the thin electrolyte layer condition.

Oxygen reduction was inhibited for corrosion product formed on Ni containing steel;

Concentration of Ni in the rust layer formed on Ni containing steel may led to the formation of amorphous products;

Ni distributed in the structure of Fe_3O_4 in the rust formed on Ni containing steel, thereby leading to the formation of amorphous products,thisresults in the inhibition of oxygen reduction during corrosion.

Reference

[1] R.E.Melchers2008*Corros.Sci.* **50**3445
[2] T.Kamimura, S.Hara, H.Miyuki, M.Yamashita, H.Uchida 2006 *Corros.Sci.***48**2799
[3] X.Zhang, S.Yang, W.Zhang, H.Guo, X.He 2014*Corros.Sci.***82**165
[4] Y.Ma, Y.Li, F.Wang 2009 *Corros.Sci.* **51**1725
[5] Y.Yuan, L.Liang, C.Wang, Y.Zhu 2010 *Electrochem.commun.***12**1804
[6] C.Plennevaux, J.Kittel, M.Frégonèse, B.Normand, F.Ropital, F.Grosjean, T.Cassagne, 2013 *Electrochem.commun.***26**17
[7] M.Morcillo, B.Chico, I.Diaz, D.de la Fuente2013 *Corros,Sci.***77**6
[8] I.Sugimoto, K.Kita 2010 QR of RTRI**51**33
[9] M.Kimura, H.Kihira, N.Ohta, M.Hashimoto, T.Senuma 2005 *Corro.Sci.***47** 2499
[10] M.Itagaki, R.Nozue, K.Watanabe, H.Katayama, K.Noda 2004 *Corros.Sci.* **46**1301
[11] X.Chen, J.Dong, E.Han, W.Ke 2007*Mater.Lett***61**4050
[12] I.Diaz, H.Cano, D.de la Fuente, B. Chico, J.M.Vega, M.Morcillo2013 *Corros.Sci.* **76**348
[13] H.Cano, D.Neff, M.Morcillo, P.Dillmann, I.Diaz, D.de la Fuente 2014 *Corros.Sci.***87**438
[14] A.Nishikata, F.Suzuki, T. Tsuru 2005 *Corros.Sci.***47**2578
[15] T.Nishimura, H.Katayama, K.Noda, T.Kodama 2000 *Corros.Sci.***42**1611
[16] X.Cheng, Z.Jin, M.Liu, X.Li 2017*Corros.Sci.***115**135
[17] X.Zhong, G.Zhang, Y.Qiu, Z.Chen, X.Guo 2013*Corros.Sci.* **74**71
[18] X.Zhong, G.Zhang, X.Guo 2015 *Corros.Sci.* 961
[19] P.Yi, K.Xiao, K.Ding, G.Li, C.Dong, X.Li2016 *Trans.Nonferrous Met.Soc.*China**26**1146
[20] M.Judith, R.Solenn, F.Eddy, T.Denis, M. François 2014*Corros.Sci.***78**293
[21] S.Hœrlé, F.Mazaudier, Ph. Dillmann, G.Santarini 2004*Corros.Sci.***46**1431
[22] M.Stratmann, K.Bohnenkamp, H.J.Engell 1983*Corro.Sci.***23**969
[23] M.Stratmann, K.Hoffmann 1989*Corro.Sci.***29**1329
[24] U.R.Evans, C.A.J.Taylor 1972 *Corro.Sci.***12**227

AUTHOR INDEX